University of California

GEOLOGICAL SCIENCES
Volume 127

New Pleistocene Conifer Records, Coastal California

by Daniel I. Axelrod

University of California Press

NEW PLEISTOCENE CONIFER RECORDS, COASTAL CALIFORNIA

New Pleistocene Conifer Records, Coastal California

by Daniel I. Axelrod

UNIVERSITY OF CALIFORNIA PRESS
Berkeley • Los Angeles • London

University of California Publications in Geological Sciences

Editorial Board: D. I. Axelrod, W. B. N. Berry, R. L. Hay,
   M. A. Murphy, J. W. Schopf, W. S. Wise, M. O. Woodburne

Volume 127

Issue Date: November 1983

University of California Press
Berkeley and Los Angeles
California

University of California Press, Ltd.
London, England

ISBN 0-520-09707-6
Library of Congress Catalog Card Number: 83-6874

Library of Congress Cataloging in Publication Data

Axelrod, Daniel I.
   New Pleistocene conifer records, coastal California

   (University of California publications in geological
sciences; v. 127)
   Includes index.
   1. Conifers, Fossil. 2. Paleobotany-Pleistocene.
3. Paleobotany--California. I. Title. II. Series.
QE977.A93 1983   561'.52'09794   83-6874
ISBN 0-520-09707-6

Contents

(A detailed table of contents
precedes each chapter)

Acknowledgments

Fieldwork necessary to collect the fossil conifers described in this report, as well as suites of cones of living coastal pines for comparison with them, was carried out with funds from the Committee on Research, University of California, Davis, and National Science Foundation Grant DEB 80-25525, for which the writer is grateful.

A Gowen cypress cone was found by Leuren Moret at the Drakes Bay site where numerous cones of Monterey pine have been recovered over a period of 15 years. Later collecting at this site added two more cypress cones to the sample.

The rich deposit of conifer cones on the coast near Seacliff was discovered by Martha Coats of Santa Paula, who kindly brought the locality to my attention and generously donated to the type-collection critical specimens that she had secured. Scott Miller and Clifford Smith, staff members at the Santa Barbara Museum of Natural History, assisted in collecting and helped in other ways as well. J. Robert Haller kindly secured additional cones at the site which was exposed again by the major winter storms of 1983.

Discussions with Eugene Begg, William B. Critchfield, Donald Kyhos, Connie Millar, and Peter H. Raven of systematic and evolutionary problems raised by the coastal pines have helped me to arrive at a better understanding of their probable history. Further clarification will depend primarily on obtaining new suites of cones from both older and younger rocks.

Abstract

This volume reports on new finds of Pleistocene conifers at two sites in coastal California, providing further insight into diverse ecological, distributional, and evolutionary problems.

Ovulate cones of Cupressus goveniana occur with abundant fossil cones of Pinus radiata in Pleistocene rocks at Drakes Bay. Geologic, paleobotanic, and climatic data suggest that this cypress was restricted southward by a colder climate during the last glacial, whereas it was eliminated from its southern California occurrences by a warmer, drier post-glacial climate. The relict occurrence of Gowen cypress and other paleoendemics in the Monterey area reflects the moderating influence on near-shore climate of the Monterey submarine canyon, which provided a foggy local haven for survival during the Xerothermic.

Ovulate cones of seven fossil conifers occur in a debris-raft transported as a turbidite into the early Pleistocene (1.0 m.y.) marine upper Pico Formation near Seacliff, Ventura County. Cones of Pinus attenuata, P. muricata var. muricata, P. muricata var. stantonii, P. radiata, and Pseudotsuga menziesii indicate higher precipitation and lower temperature in the coastal strip than at present; Pinus remorata suggests well-drained, drier sites near the coast, and P. sabiniana implies drier, less equable conditions inland. Increased precipitation and lowered temperature enabled Pseudotsuga to invade coastal southern California and overlap the other conifers. The present more restricted distributions of all these taxa developed as the climate became drier and warmer during the Xerothermic. Two new Pinus radiata populations (Año Nuevo and Cambria) and hybrid populations of P. muricata X remorata appear to have originated in response to it. The occurrence of six pine species of varied ecologic and climatic requirements in the Seacliff deposit points up the limitations of fossil-pollen studies that identify Pinus only to the level of genus.

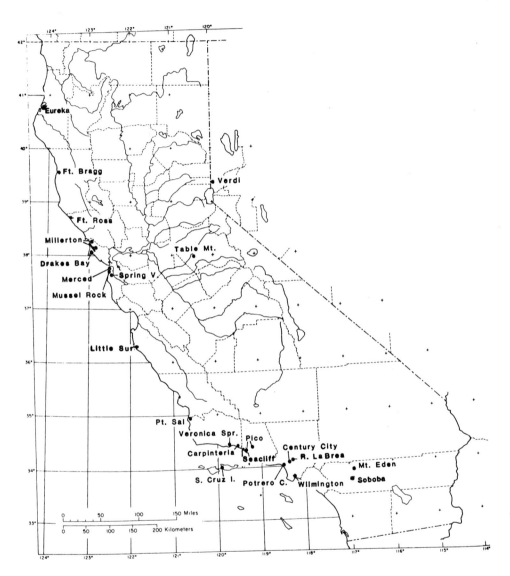

FIGURE 1. Principal localities to which reference is made.

CHAPTER I

GOWEN CYPRESS AT DRAKES BAY, MARIN COUNTY

Contents

INTRODUCTION

Although most of the 10 living species of Cupressus in California have very restricted distributions (see Griffin and Critchfield, 1972), the fossil record shows that some of them were distributed more widely in the past. In California, four species have been described from the Tertiary and three from the Quaternary (Table 1). Most of the Tertiary species are known from only one or two specimens, the chief exception being C. mokelumensis (Axelrod, 1980b). In the Quaternary, Rancho La Brea (Pit 91) provides the richest record of fossil cypress (Warter, 1976), with three species reported: C. forbesii Jepson, C. goveniana Gordon, and C. macrocarpa Hartweg. Of these, Monterey cypress (C. macrocarpa) is known from fully 300 cones deposited close to the trunk of a large tree, and numerous branchlets and seeds have also been recovered (Warter, 1976).

A new record of C. goveniana is represented by the recovery of three ovulate cones at the Drakes Bay locality that earlier yielded numerous cones of Pinus radiata similar to those produced by the modern population at Monterey-Carmel (Axelrod, 1980a, plates 8-10). This new record of cypress is 200 km north of its present area, which comprises only the small stands in the vicinity of Monterey and Gibson Creek (Fig. 2). Geologic, floristic, and climatic data provide a basis for inferring the historical factors that appear to account for its present restricted distribution.

TABLE I

Records of fossil species of <u>Cupressus</u> in California

SPECIES AND REFERENCES	MODERN ANALOGUES	FOSSIL STRUCTURES	OCCURRENCE AND AGE
Tertiary			
C. forbesii[1]	forbesii	ovulate cone, branchlet	Mt. Eden, 5 m.y.
C. mohavensis[2]	arizonica or nevadensis	ovulate cones, branchlets	Tehachapi, 17 m.y.
C. mokelumensis[3]	funebris	ovulate cones, branchlets,	Mt. Reba, 7 m.y.
C. ricardensis[4]	nevadensis or forbesii?	wood	Ricardo, 12 m.y.
Quaternary			
C. forbesii[5]	forbesii	ovulate cones	Rancho La Brea, 30,000 B.P.
C. goveniana[5]	goveniana	ovulate cones	Rancho La Brea, 30,000 B.P.
C. goveniana[6]	goveniana	ovulate cones, wood, branchlets	Santa Cruz I., 15,000 B.P.
C. goveniana[7]	goveniana	ovulate and staminate cones, leafy twigs	Carpinteria, 40,000 B.P.
C. goveniana[8]	goveniana	ovulate and staminate cones, leafy twigs, seeds, wood	Millerton, 28,000 B.P.
C. macrocarpa[5]	macrocarpa	ovulate cones, seeds, leafy branchlets, wood	Rancho La Brea, 30,000 B.P.

1-Axelrod, 1937; 2-Axelrod, 1939; 3-Axelrod, 1980b; 4-Webber, 1933; 5-Warter, 1976; 6-Chaney and Mason, 1930; 7-Chaney and Mason, 1933, 8-Mason, 1934.

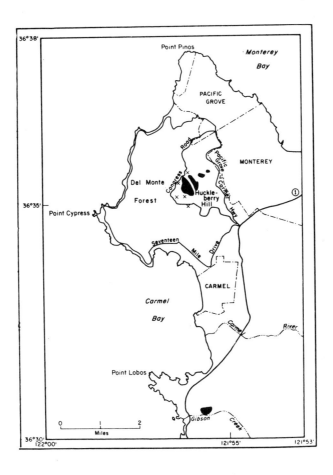

FIGURE 2. Present occurrence of <u>Cupressus goveniana</u>
(from Griffin and Critchfield, 1972, map 22). Black areas
represent stands, x's scattered trees.

OCCURRENCE AND AGE

The <u>Cupressus goveniana</u> cones found at Drakes Bay were
preserved in a buff, lenticular siltstone which rests uncon-
formably on the upper Miocene Drakes Bay Formation (Axelrod,
1980a), whose base is dated at 9.3 $\pm$ 0.5 m.y. (Galloway,
1977). Exposed in a low seacliff (Pl. 1, fig. 1), the
siltstone is about 1.5 m thick near the center of the out-
crop; it lenses out laterally within a distance of scarcely

75 m and is covered upslope by stabilized dunes. Numerous disoriented sticks, branches, and perfectly preserved ovulate cones of Pinus radiata are scattered through the siltstone. The deposit is a turbidite, with the silty matrix derived chiefly from erosion of the underlying Drakes Bay Formation. The Drakes Bay rests on the Monterey Formation, which laps onto a quartz diorite basement; rare, rounded pebbles of chert and quartz diorite from the latter formations are also in the turbidite.

The siltstone has numerous angular to subangular clasts of a very fine tuff similar to the vitric tuffs in the Sonoma Volcanics which yielded an age of 3.5 m.y. at the Petrified Forest east of Santa Rosa (Evernden and James, 1964). This was earlier considered the age of the cone deposit (Axelrod, 1980a). However, it is now known that generally similar tuffs occur widely in this part of coast-central California, ranging in age from 7.0 to 0.5 m.y. (Sarna-Wojcicki, 1976, 1979; Mathieson and Sarna-Wojcicki, 1982). Since this casts doubt on the earlier presumed age of the Drakes Bay deposit, a sample of the tuff was submitted for radiometric dating. The laboratory reports that it is so contaminated with fine debris that a reliable age cannot be established. A radiocarbon date on the wood in the deposit indicates that it is older than 36,480 B.P. (R. E. Taylor, Jr., Univ. California, Riverside, Radiocarbon Laboratory, December 1981). In view of the uncertainty as to age, a tentative date is assigned to the deposit by using the presumed rate of evolution of cone size suggested by Pinus radiata records, whose ages are reasonably well established. Since cone size appears to have increased gradually, and was generally paralleled by seed-protein differences in these populations (Murphy, 1981), an approximate age of 0.5 m.y. is inferred for the pine and associated cypress cones.

PALEOECOLOGY

Vegetation

Only <u>Pinus</u> <u>radiata</u> cones have been recovered previously
from the turbidite that yielded the cypress cones. The
absence of other species (apart from unidentified woods) in
the deposit is not surprising. As the heavier cones and
sticks settled from suspension, the lightweight seeds and
other small structures bypassed them, settling out at a more
distant site. The rarity of cypress cones as compared with
more than 70 pine cones suggests that cypress cones either
were transported farther offshore, or that cypress was
less abundant than pine in the bordering hills, or was
confined to more localized habitats. And since the small
trees retain cones for long periods, this may also account
for their rarity in the deposit. In any event it is clear
that a pine forest largely covered the shore area. Cypress
may have been confined chiefly to low hills of quartz
diorite, then at the site of present Inverness Ridge, 5 km
east. The pines probably occurred there as well as on heavy
soils derived from the Monterey and Drakes Bay Formations
that skirted the front of the low granitic ridge, as shown on
the geologic map (Galloway, 1977). Such a setting parallels
the occurrence of these conifers at Monterey today (Pl. 1,
fig. 2), for the cypress groves are confined to quartz
diorite at Huckleberry Hill and on Gibson Creek, 9 km south
(Fig. 2), whereas the pine occurs on granitics as well as on
the more widely distributed siliceous shales of the Monterey
Formation (see Jennings and Strand, 1958).

Climate

Climatic change accounts for the restriction of <u>Cupressus</u>
<u>goveniana</u> to its present area 200 km southeast. Inasmuch as
precipitation is generally similar along the coast, ranging
from 500 to 760 mm or more, depending chiefly on latitude and

coastal terrain, it probably was not critical in limiting
cypress distribution except during the Xerothermic (see
below). As for the temperature, three lines of evidence
suggest that climate during the last glacial age (20,000-
15,000 B.P.) was probably too cold for Gowen cypress in the
Drakes Bay area.

Geologic Evidence

Glacial features (moraines, hanging valleys, striated
and polished pavements, U-valleys) indicate that mountain
glaciers extended down valleys to middle elevations in the
Sierra Nevada, Trinity Mountains, and Klamath Mountains
(Flint, 1971), implying lower temperatures everywhere in the
region. At that time, with water tied up in ice, sea level
was lower than at present. Along the central California
coast the shoreline was fully 40-50 km farther west (Helley
and LaJoie, 1979, fig. 12) during the late Wisconsin (15,000
B.P.), out near the present edge of the continental shelf.
Upwelling of frigid, deep bathyal to abyssal water from the
continental slope, which was then at or close to the shore,
would result in a colder maritime climate than now exists in
central California, certainly one more extreme than that
under which C. goveniana now lives.

The fossil locality, 9 km west of the San Andreas rift,
has been moving northwesterly into a cooler climate. With an
average rate of movement of 5 cm/yr, the inferred age of the
flora (0.5 m.y.) indicates that the locality would have been
about 25 km southeast. In terms of present geography, this
would place it on the outer coast in the vicinity of Bolinas.
Mean annual temperature there probably is about 1°C higher
than that now at the fossil site, for it lies 24 km farther
west, jutting out into the ocean.

Paleobotanic Evidence

Colder climate in the Drakes Bay area during the late
Pleistocene is implied by the occurrence of Picea sitchensis

in the Millerton flora (Mason, 1934), situated 10 km north-
east and dated at 29,000 years B.P. (Berger and Libby, 1966).
At that time, spruce trees probably grew on nearby Inverness
Ridge (Axelrod, 1980a, p. 11), 2.5 km southwest of Millerton
and 7 km east of the Drakes Bay site. The principal southern
stands of Picea are now near Ferndale, 300 km north of Drakes
Bay. From that area there is a 140 km gap in its distribu-
tion to a few relict groves between Fort Bragg and Big River
(Griffin and Critchfield, 1972), about 160 km north of Drakes
Bay (Fig. 3). P. sitchensis has also been reported from

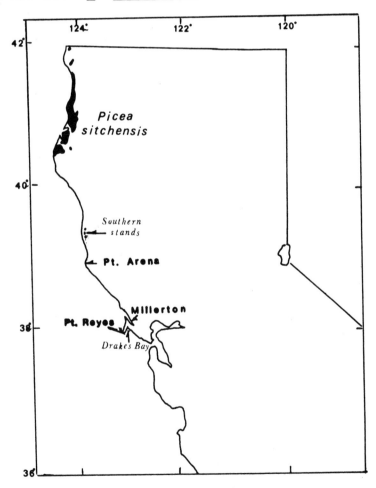

FIGURE 3. Modern distribution of Picea sitchensis on the
northwest coast of California (from Griffin and Critchfield,
1972, map 39). Late Pleistocene sites for Picea are at
Point Arena and Millerton.

the late Pleistocene at Point Arena (Mason, 1934, p. 107), 45 km south of the present southernmost relict stands and 115 km north of Drakes Bay. Another indication of cooler climate in the recent past is the occurrence in the Inverness area of taxa that are at or near their southern limit of distribution, notably Ledum glandulosum, Rubus spectabilis, and Sambucus callicarpa. Furthermore, Pinus muricata represents the var. borealis which is found farther north; it certainly is quite different from the stand on Huckleberry Hill, Monterey, which is not so robust, more open, and has more heavily armed cones.

The inferred southward retreat of cypress during the last glaciation (20,000-15,000 B.P.) is consistent with the colder climate implied by the record of montane conifer forest at sea level near Mountain View, at the southwest corner of San Francisco Bay, at 23,000 B.P. (Helley et al., 1972). The megafossil flora includes species of Calocedrus, Cupressus, Pinus, and Pseudotsuga. The assemblage is confined now to higher, cooler, and moister sites in the Coast Ranges to the north, whereas the present site is in oak-grassland vegetation. The difference in implied thermal conditions is on the order of a mean annual temperature about 3°C lower than that now at the site (Axelrod, 1981, p. 850). Whereas the area would have regularly received light winter snow, it is essentially unknown there today. Although the Mountain View locality is separated from the coastal strip by the Santa Cruz Mountains, a colder climate no doubt also affected the coastal region. As noted above, during the last glaciation sea level was lower and the Pacific shore was farther west (Helley and LaJoie, 1979, fig. 12). At that time the Drakes Bay site was 40 km inland from the shore and would have had a climate much more rigorous than that at Monterey today.

Thermal Evidence

Temperature conditions that probably account for the southward restriction of cypress can be estimated by two

methods presented by Bailey (1960, 1964). First, we can com-
pare mean temperatures of the warmest month (tw) and coldest
month (tc). At the fossil locality, situated between Point
Reyes (lighthouse) and Point Reyes Station (village) (Fig.
4), mean temperature of the warmest month (tw 13.0°C) is
nearly 4°C lower than at Monterey (tw 16.8°). By contrast,

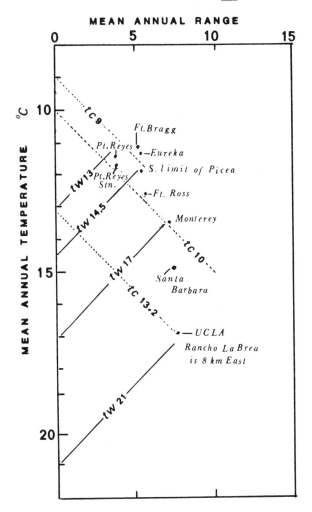

FIGURE 4. Mean temperature of the warmest month (tw)
and coldest month (tc) at stations that provide data regarding the thermal parameters of Cupressus goveniana,
past and present.

cold-month temperatures (tc 9.8°C) at the fossil locality and
Monterey are similar. However, the cold month at the fossil
site is 0.6°C warmer than that in the southern relict stands
of Picea today (tc 9.2°C), located between Fort Ross and Fort
Bragg, and it is about 1.2°C warmer than that at Eureka (tc
8.6°C), in the principal southern area of Picea.

The data suggest that during the last glacial stage,
which evidently corresponds to the Tioga glaciation in the
Sierra Nevada (R. Shaw, personal communication, January
1982), climate became too cold for Gowen cypress in the
Drakes Bay area. It must be recalled that the occurrence of
Picea sitchensis, together with Arctostaphylos uva-ursi and
Montia howellii, in the preglacial Millerton flora, which is
associated with a shallow warm-water marine molluscan fauna
(Mason, 1934; Campbell, 1974), is not anomalous (Axelrod,
1980a, pp. 11-13). The plants with generally cold requirements occur in fluvatile deposits filled with granitic debris
derived from Inverness Ridge, 2.0-2.5 km west and southwest.
These taxa were probably relictual from an earlier glacial
age, surviving in cool, northeasterly-facing canyons of the
Ridge at elevations up to 300-350 m above the site where the
flora accumulated. The marine fauna and the associated small
Guadalupe Island-type cones of Pinus radiata that dominate
the flora both imply a relatively mild climate which reflects
warmer water in the protected bay.

Mean temperatures of the warmest and coldest months provide only a first approximation of the thermal parameters of
the area occupied by a species, or a vegetation zone. They
do not indicate either (1) the warmth ($\underline{W}$) of climate, which
provides a measure of the length of the growing season; (2)
the equability ($\underline{M}$) of climate, which gives an indication of
thermal extremes; or (3) frost frequency (see Bailey, 1960,
1964). These can be determined by plotting mean annual tem-
perature ($\underline{T}$) and mean annual range of temperature ($\underline{A}$) on

the Bailey monogram (Fig. 5).   At the Drakes Bay locality, situated between Point Reyes Station (village) and Point Reyes (lighthouse), warmth of climate (W̲ 12.3°C, or 132 days with mean temperature warmer than 12.3°C) is lower than at Monterey (W̲ 13.7°C, or 174 days warmer than 13.7°C). All these stations are characterized by high equability (M̲ 70 or more; read the arcs, Fig. 5) and are well removed from regular frost, even though mean cold-month temperature is quite low (Fig. 4, dotted lines).   These data imply that

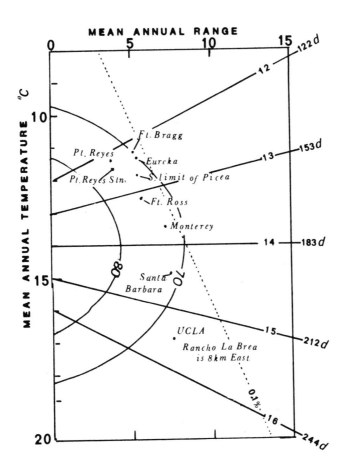

FIGURE 5.   Warmth (W̲) and equability (M̲) at stations that provide data on present and past conditions in areas of *Cupressus* *goveniana* occurrence.

Cupressus goveniana is no longer native to the Drakes Bay area because there is now, and was in the recent past, a cooler, shorter growing season than that now at Monterey.

C. goveniana has been recorded also at three sites in southern California: Santa Cruz Island (Chaney and Mason, 1930), Carpinteria (Chaney and Mason, 1933), and Rancho La Brea (Warter, 1976). The former two sites are at sea level and near Santa Barbara. Thermal conditions at Santa Barbara airport, situated close to the shore, differ markedly from those implied by the records of fossil cypress in the nearby area. Mean temperature of the warm month (tw) is 18.4°C, compared with 13.7°C at Monterey; mean temperature of the cold month (tc) at Santa Barbara Airport is 10.8°C, compared with 9.8°C at Monterey. Clearly, the warmer climate now at Santa Barbara, and presumably more so during the Xerothermic, may be the critical factor accounting for the absence of Cupressus in that area today. Warmth of climate on the Santa Barbara coast is now W 14.5°C (197 days warmer than 14.5°C) and the area is essentially frostless, with an equability rating of M 72. This compares with W 13.7°C at Monterey (174 days warmer than 13.7°C). The data imply that C. goveniana is adapted to a cooler climate than that of the Santa Barbara region today.

This conclusion agrees with data provided by the occurrence of C. goveniana at Rancho La Brea (Warter, 1976). Fig. 5 shows that there is a much warmer climate there (W 15.5°C, or 227 days with mean temperature warmer than 15.5°C) than that in which cypress now lives (W 13.7°C, or 174 days warmer than 13.7°C). Note also that equability (read the arcs) is M 67 at Rancho La Brea (i.e., UCLA), as compared with M 74 at Monterey. However, since Rancho La Brea is 8 km inland from UCLA and has both warmer summers and cooler winters, equability there is probably nearer M 65. Clearly, the development of a warmer, less equable climate at Rancho La Brea following the last glacial would have been unsuited for C. goveniana,

as would the lower rainfall. Fig. 4 shows that mean warm-
month temperature at Rancho La Brea (tw 21°C) is 4°C higher
than that now at Monterey (tw 17°C). It would have been
higher at Rancho La Brea during the Xerothermic as judged
from interior vertebrates recorded there, notably desert
scaly lizard (Sceloporus magister), yucca night lizard
(Xantusia vigilis), and desert shrew (Notisorex) [Axelrod,
1966, p. 44].

Furthermore, Warter (1976) reports the presence of
several rare plants that are typically of inland occurrence
today, well removed from the dominant closed-cone pine
forest. Among these are Arctostaphylos glauca, A. pungens,
Juniperus californica, Pinus sabiniana, and Quercus lobata.
They may be holdovers from the preceding warmer interglacial
stage, persisting on local drier sites in the nearby hills,
or on old alluvial fans, and thus contributed only rarely to
the record. A similar relationship is shown by the Carpinteria flora (Chaney and Mason, 1933), with rare specimens of
A. glauca, J. californica, and P.sabiniana evidently carried
from nearby Eocene sandstone dip-slopes in the adjacent hills
to the coastal strip dominated by closed-cone pine forest.

Mean temperature of the cold month (tc 13.2°C) at Rancho
La Brea is 2.4°C warmer than at Monterey (tc 9.8°C; Fig. 4).
Mean annual temperature at Rancho La Brea is now 3.4°C higher
than at Monterey (Fig. 4). That it would have been even
higher during the Xerothermic is evident from a comparison of
temperatures at Rancho La Brea with those in the interior
valleys of southern California (Table 2), where taxa now
occur that are recorded at Rancho La Brea. In the interior,
mean temperature of the warmest month is 3.0°C higher and
that of the coldest month 2.4°C lower than at Rancho La Brea
today. The cold month is fully 3°C lower than in coastal
stations near the late Pleistocene sites at Carpinteria and
Santa Cruz Island. This is reflected in the lower equability

TABLE 2

Comparison of temperatures (°C) in interior valleys of
southern California with that now near Rancho La Brea

STATIONS	MEAN ANNUAL	WARMEST MONTH	COLDEST MONTH	EQUABILITY RATING
Cachuma Lake	16.0	22.2	9.8	61
Claremont	16.8	23.6	10.6	59
Ojai	16.3	22.9	10.1	60
Pasadena	17.3	23.2	11.8	60
San Antonio Canyon	17.4	24.9	10.9	57
San Bernardino	17.7	25.3	10.8	56
San Fernando	17.4	23.5	11.8	59
Mean	16.9	23.7	10.8	59
Rancho La Brea (UCLA)	16.9	20.7	13.2	67

at the interior stations (mean = M 59) as compared with
Rancho La Brea (M 67) or Santa Barbara (M 72).

DISCUSSION

The question arises as to whether Gowen cypress may have
had a wider distribution in the interior during the Quaternary. As noted above, the late Pleistocene (23,000 B.P.)
megafossil flora near Mountain View shows that when a montane
conifer forest was at sea level, mean annual temperature was
about 3°C lower than that in the area today (Axelrod, 1981)
and there was regular winter snow. This implies climate in
interior central California was much too severe for Cupressus
goveniana. This may also be inferred from the record of
valley glaciation in the Sierra Nevada, for cold-air drainage
would have rendered the interior lowlands quite cold.

In southern California, the early Pleistocene Soboba
flora (Axelrod, 1966), 140 km east-southeast of Rancho La
Brea (Fig. 1), is associated with the Bautista mammal fauna

of Irvingtonian age (Frick, 1921; Savage, 1951; Marshall et al.1982). It is about 1.0 m.y. old, contemporaneous with the Sherwin glaciation in the Sierra Nevada and Nebraskan glaciation in the mid-continent (Boellstorff, 1978). The Soboba flora shows that mixed conifer forest, represented by megafossil remains of Abies concolor (Gord. & Glend.) Lindley, Calocedrus decurrens (Torrey) Florin, Pinus lambertiana Douglas, P. ponderosa Lawson, and Populus tremuloides Michaux, reached down into an area now semidesert. This suggests that mean annual temperature at the Soboba site was about 11.1°C, as compared with 16.7°C today, and that precipitation was near 760 mm, compared with 330 mm now at the locality (Axelrod, 1976, Fig. 5). Light snow was common then but is rarely encountered in the Soboba area today. The evidence suggests that if there were any stands of Cupressus goveniana in the interior, they were also eliminated by cold during the glacial ages.

Gowen cypress probably disappeared from coastal southern California as a spreading warmer, drier Xerothermic climate facilitated the coastward movement of semidesert taxa (Axelrod, 1966, pp. 42-55). Furthermore, as this climate enabled species from the south Coast Ranges and Mohave Desert to extend northward and coastward into the south and central Coast Ranges, stands of Gowen cypress at intermediate sites along the coast would have gradually disappeared, except for those near Monterey. The rapid spread of more xeric climate also appears to have broken up the essentially continuous late Pleistocene closed-cone pine forest into the present scattered patches along the outer coast. These groves are confined to areas where summer fog is more frequent and more persistent. This relationship was especially apparent during a low-level (2,100 m) air trip to Santa Barbara made on the afternoon of July 30, 1980. Patches of fog were noted only at Año Nuevo, Monterey-Carmel, Cambria, Pecho Hills, Purisima Hills, and Santa Cruz Island, all areas that now support closed-cone pine forests. As discussed elsewhere (Axelrod, 1982), the Monterey area is unique in that it is the only site along the central coast where a major submarine canyon trenches the continental shelf to the near-shore area. A strong upwelling of cold water gives the area a higher fog frequency than elsewhere along the central coast. This more constant fog cover evidently provided the paleoendemics of the Monterey area (*Cupressus goveniana*, *C*. *macrocarpa*, *Pinus radiata*) with an effective shield from the spreading drier, warmer Xerothermic climate, and probably accounts for their persistence there.

A favorable local climate may not be the sole factor that accounts for the persistence of *C*. *goveniana* on the lower-middle west slope of Huckleberry Hill, where it makes up a pygmy forest. The cypresses are relatively small, mostly 4-5 m tall. The tallest one noted was about 12-13 m tall and occurs with *Pinus radiata* somewhat removed from the main cypress grove. The cypress together with *P*. *muricata* occupies a relatively flat, gently sloping (2-8°) terrain that evidently represents an ancient elevated terrace. The trees occur on a strongly podsolized soil. The light gray or white loamy fine-sand surface soil of the Narlon Series is about 1 ft thick and overlies a claypan subsoil up to 2.5 ft thick. The claypan is buff to yellow-brown, locally rust-mottled, and highly acid in reaction (Cook, 1978). Elsewhere in the area there is a cemented sand at a depth of 10-20 in, rather than a claypan. Runoff is slow to medium, and temporary shallow ponds form in swales in wet winters.

In addition to *P*. *muricata*, the pygmy forest area has several *P*. *remorata* trees. These appear to be relict, surviving on a site where water availability is low, since the roots cannot penetrate the claypan. By contrast, *P*. *remorata* is quite rare in the *muricata* stand higher up on steeper slopes adjacent to the Carmel-Pacific Grove Highway. The *muricata* trees are fuller in foliage and not so stunted at this higher elevation, where precipitation probably exceeds 635 mm (25 in) and there is less soil-profile development. P. radiata occurs only as scattered, rare trees on the gently sloping erosion surface that supports the pygmy forest, but dominates the steeper, well-drained slopes covered with the slightly acid Sheridan soil derived from the granitics. It is apparent that C. goveniana, P. muricata, and P. remorata cannot compete as well with P. radiata on the moister, steeper, well-drained slopes. That competition is another factor accounting for present restriction of the cypress to poorly-drained podsolic soil is implied also by the restriction of a pygmy forest on the Mendocino Coast to similar sites.

Dwarf conifers (P. bolanderi, C. pygmaea) and associated heaths (Arctostaphylos, Gaultheria, Ledum, Vaccinium) are well developed in the Fort Bragg region. Their occurrence, described by Gardner and Bradshaw (1954) and by Jenny et al.(1961), is restricted to a ground-water podsol designated the Blacklock Series. As Jenny et al. note, "on level plateau surfaces with high and fluctuating water-tables, the podsolic processes are intensified, resulting in hardpans and claypans and soil conditions that produce dwarfism, narrow endemics, and ecotypic differentiation." The pine occurs in cane-like stands 6-15 decimeters tall and has very small cones. The cypress also forms dense stands, usually a little over 1-2 m tall. As they move away from the bogs both trees increase rapidly in stature, with the cypress reaching a height of 40 m. It appears that both are persisting in the poorly-drained sites because they escape competition from trees of the bordering dense forest composed of Abies, Pseudotsuga, Sequoia, and Tsuga, a community of which they are scarcely a part.

The Blacklock soil on which the dwarf taxa occur in coastal Mendocino County is also in the Marshfield area, coastal southern Oregon. The question thus arises as to the conditions, other than poor drainage, that may have produced the podsolic soil on Huckleberry Hill which supports the pygmy forest there. A cooler, moister climate probably accounts for its development. This is suggested by several plants in the Monterey area that are now at or near their southern limit of distribution, notably Arctostaphylos uva-ursi, Rhododendron macrophyllum, and Xerophyllum tenax. These are evidently remnants of a flora that entered the area when its climate was more like that now well to the north. A larger number of taxa of this alliance are in the Santa Cruz Mountains directly north. Thomas (1961) noted that 160-odd taxa reach their southern distribution there, including Ceanothus foliosus, C. velutinus var. laevigata, Euonymus occidentalis, Fraxinus latifolius, Prunus subcordata, Rubus leucodermis, R. spectabilis, and Vaccinium parvifolium, as well as the northern species in the Monterey area.

That these taxa of northern affinity probably entered the region during the cooler, moister phases of the Pleistocene is implied by the nature of the Pleistocene floras in the coastal sector. As noted earlier, at Tomales Bay, the Millerton flora (Mason, 1934) includes Picea sitchensis, Camassia leichtlinii, and Montia howellii, which now occur farther north in a cooler coastal climate. In addition, the San Bruno flora south of San Francisco (Potbury, 1932), which is early postglacial, suggests a climate like that at Inverness to the north. In coastal southern California, the Santa Cruz Island (Chaney and Mason, 1930) and Carpinteria (Chaney and Mason, 1933) floras both indicate conditions comparable to those now at Monterey. This implies a southward shift of cooler, moister climate on the order of 320-400 km (200-250 mi), though it may have been somewhat greater. The Fort Bragg area where the pygmy forest is well developed, is 320-400 km north of Monterey. The data suggest that podsolic soil at Huckleberry Hill may have developed when conditions, at a minimum, were like those now in the Fort Bragg region (Table 3). This implies a mean annual temperature about 2°C

TABLE 3

Climatic data for estimating podsolic soil formation in the
pygmy forest area, Huckleberry Hill, Monterey

STATIONS	TEMPERATURE MEAN ANN.	RANGE[a]	WARMTH[b]		PRECIPITATION	
Bandon, Ore.	10.8°C	6.4°C	12.2°C	129 days	1,400 mm	55 in
Fort Bragg, Calif.	11.6	4.9	12.5	138	1,000	40
Monterey, Calif.	13.5	7.0	13.9	179	635[c]	25

[a]Range of temperature=difference between mean temperature of warmest and coldest months.

[b]Warmth refers to the number of days warmer than the stipulated temperature (Bailey, 1960).

[c]Monterey at sea level has 450 mm (17 in) rainfall but cypress lives on the slopes of Huckleberry Hill where rainfall is certainly higher. It may be more than 635 mm (25 in), perhaps as much as 760 mm (30 in), but probably not less than 635 mm.

lower and a rainfall at least 380 mm (15 in) higher than that now at Monterey. These estimates are consistent not only with the latitudinal shift of climate implied by Pleistocene floras in the coastal sector, but with the general climatic requirements indicated by the numerous plants in the Santa Cruz Mountains that find optimum development in the coastal sector well to the north (Thomas, 1961).

SUMMARY

Rare ovulate cones of Cupressus goveniana found in Pleistocene rocks at Drakes Bay are associated with Pinus radiata cones. Cypress may have occupied quartz diorite outcrops along Inverness Ridge 5 km east, whereas the abundant pine probably occurred there and on shales of the Monterey and Drakes Bay Formations that lapped onto the granitics, a distribution similar to their occurrence at Monterey today.

The cypress was probably restricted southward during the last glacial by a climate of less warmth. With lower sea level, the continental shelf was emerged some 40-45 km farther west, bringing a colder climate enhanced by an upwelling of cold, abyssal water from canyons trenching the continental slope which then bordered the shore. Colder climate is indicated also by mixed conifer forest that covered lowlands of the San Francisco Bay area, implying a mean annual temperature about 3°C lower than at present. Records of spruce in late Pleistocene coastal sites south of its present area also imply colder climate. Temperature data for stations along the coast show lower warmth in the Drakes Bay area today and in the past than at Monterey, where the cypress now lives.

Cypress evidently disappeared from coastal southern California (Rancho La Brea, Carpinteria, Santa Cruz Island) as spreading Xerothermic climate (8,000-4,000 B.P.) eliminated it there as well as at intermediate sites along the coast, except for the relict groves at Monterey. There a more persistent fog blanket, resulting from upwelling from the Monterey-Carmel submarine canyon, evidently shielded it and other paleoendemics from the drier, warmer Xerothermic climate.

Ecologic evidence suggests that the cypress at Huckleberry Hill survives on a relatively sterile podsolic soil chiefly because it and the associated Pinus muricata escape competition there from P. radiata. Competition with dense coastal conifer-forest taxa may also explain the restriction of C. pygmaea and C. bolanderi to similar soils on the Mendocino coast. On sites away from podsolic soils of poor drainage, these dwarf ecotypes merge gradually into normal forest trees.

SYSTEMATIC DESCRIPTIONS

All specimens are deposited in the
Museum of Paleontology, University of California,
Berkeley.

Family PINACEAE
Pinus radiata D. Don
(Plate 2, figs. 4-6; Pl. 3, figs. 1-3)
Pinus radiata D. Don, Linn. Soc. London Trans.17, p. 441,
1836.
Mason, Carnegie Inst. Wash. Publ. 346, p. 147, pl. 1,
fig. 2, 1927.
Chaney and Mason, Carnegie Inst. Wash. Publ. 415, p. 55,
pl. 4, fig. 2; pl. 5, figs. 4, 9, 1933.
Axelrod, Univ. Calif. Publ. Geol. Sci. 120, p. 5, pls.
4-10, 1980.
As outlined earlier, cones of P. radiata are perfectly
preserved in the Drakes Bay deposit. Several additional pine
cones were collected during the search for cypress cones and
are figured here. They are well within the mean size of
those produced by the present Monterey population.

One of the cotypes (no. 307) of Pinus masonii Dorf
(1930, pl. 5, fig. 6) is allied to P. radiata, not P.
muricata. The apophyses are rounded above and also bluntly
acute, features typical of numerous cones in the Monterey
population.

In an earlier report (Axelrod, 1980a, pp. 9-10), concern
was expressed regarding the large size of the only P. radiata
pictured by Mason (1934) from the Millerton flora at Tomales

Bay. Additional evidence now clearly shows that that cone is
in fact from the Drakes Bay, not the Millerton site. Typical
cones from the Millerton site are markedly smaller and the
apophyses are not so prominently developed (Axelrod, 1980a,
Pls. 2, 3) as in those from Drakes Bay (see Pls. 2 and 3;
also Axelrod, 1980a, Pls. 8, 9, 10).

    <u>Collection</u>: Hypotypes nos. 6300-6305; homeotypes nos.
    7136-7142.

Family CUPRESSACEAE
<u>Cypressus goveniana</u> Gordon
(Plate 2, figs. 1-3)

<u>Cupressus goveniana</u> Gordon, Jour. Hort. Soc. London 4,
    p. 295, 1849.
    Chaney and Mason, Carnegie Inst. Wash. Publ. 415, p. 10,
        pl. 5, figs. 4-6; pl. 7, figs. 1-9, 1930.
    Chaney and Mason, Carnegie Inst. Wash. Publ. 415, p. 58,
        pl. 6, figs. 1-5, 7, 8, 1933.
    Mason, Carnegie Inst. Wash. Publ. 415, p. 149, pl. 6,
        fig. 1, pl. 7, figs. 5-9, 1934.

The Drakes Bay deposit has yielded three ovulate cones
that are ovoid to slightly oblong as a result of compression.
They range from 25 to 28 mm long and 23 to 25 mm wide, and 8
to 10 cone scales are oppositely arranged. The outer surface
of the scales is smooth; the surface of one cone is partly
decomposed and hence appears striated because the vasculature
is exposed. The smooth scales have no prominent apophyses
and the umbos are crescent-shaped. The cones were open at
the time of burial, so seeds were not recovered.

    The dimensions of the cones, the absence of apophyses
on the scales, and the crescentic umbos on the smooth scales,
stamp them as allied to <u>C</u>. <u>goveniana</u>, a relict now known only
from the vicinity of Monterey. Its fossil records include
sites at Tomales Bay, Carpinteria, Santa Cruz Island, and
Rancho La Brea, all of late Pleistocene age.

    <u>Collection</u>: Hypotypes nos. 6297-6299.

CHAPTER I PLATES

PLATE 1

Drakes Bay fossil site and allied modern vegetation

Fig. 1.   The Drakes Bay site is in a low seacliff.   In
addition to cones of Cupressus goveniana, this site yields
numerous cones of Pinus radiata like those of the living
Monterey population.   A P. muricata forest covers the distant
slopes of Inverness Ridge and gives way to a Pseudotsuga
forest in the higher, moister part of the range to the south
(right), where Monterey shale covers the quartz diorite.

Fig. 2.   Cupressus goveniana (right) and Pinus radiata on
the middle, west slope of Huckleberry Hill, Del Monte Forest,
Monterey, California.

PLATE 2

Drakes Bay fossils

Figs. 1-3. Cupressus goveniana Gordon. Hypotypes nos. 6297-6299. The white angular clasts in the matrix of Figure 3 are dacitic tuffs.

Figs. 4-6. Pinus radiata D. Don. Hypotypes nos. 6300-6302.

PLATE 3

Drakes Bay fossils

Figs. 1-3. Pinus radiata D. Don. Hypotypes nos.
6303-6305.

CHAPTER II

A CONIFER FOREST NEAR SEACLIFF, VENTURA COUNTY

Contents

Figures

Tables

Plates

INTRODUCTION

In view of the intense tectonic activity in California during the Quaternary (Bailey and Jahns, 1954; Christensen, 1966), one might suppose that sediment pouring off rapidly rising mountains into the numerous basins of deposition would have preserved an exceedingly rich record of land life. But apart from a few tar-pit deposits where large biotic samples have been recovered, as at Rancho La Brea, Carpinteria, and McKittrick, few rich samples of megafossil floras or mammal faunas have been found. The fossil floras occur chiefly in the coastal strip (Fig. 1): at Millerton (Mason, 1934), San Bruno (Potbury, 1932), Little Sur (Langenheim and Durham, 1963), Carpinteria (Chaney and Mason, 1933), Santa Cruz Island (Chaney and Mason, 1930) and Rancho La Brea (Warter, 1976). Most of them are of late Wisconsin age, largely in the range of from 30,000 to 12,000 years ago.

The recent discovery of perfectly preserved ovulate cones of seven conifers in a deposit in coastal southern California represents the first reasonably rich coastal-strip plant record from the early Pleistocene. This flora from near Seacliff (Pl. 4, fig. 1), 15 km west of Ventura, is closely associated with a fine vitric tuff ("Bailey tuff") that has been dated radiometrically at 1.2 m.y., thus enhancing the chronologic and evolutionary importance of the species. The Seacliff flora represents an entirely different environment from that indicated by the contemporaneous

Soboba flora at the base of the San Jacinto Mountains, 235
km southeast in interior southern California (Axelrod, 1966).
These floras thus provide a reliable basis for reconstructing
the regional distribution of vegetation in southern California at a time contemporaneous with Sherwin glaciation in the
Sierra Nevada and Nebraskan glaciation in the mid-continent.

OCCURRENCE AND AGE

The Seacliff site was exposed in the 1979-80 winter season by major storm waves that eroded the sand beach back
sufficiently to uncover the cone-bearing beds which are below
mean sea level and accessible for only a brief time at minus
tide. By late spring (May 1980), waves had carried sand back
to cover the plant-bearing strata. Another large sample was
secured by Dr. J. Robert Haller in the 1983 winter season
when gigantic storm waves reopened the site.

The fossil locality is close to the crest of the Rincon
Anticline (Jennings and Troxel, 1954, fig. 17). Two gently
dipping beds that contain cones and associated branches are
about 20 cm thick and 1 m apart stratigraphically. The lower
few centimeters are coarse sandstone and grit with scattered
well-rounded pebbles grading up into the plant debris which
is encased in a finer pebbly sandstone. Plant material makes
up the bulk of the upper 5-7 cm of each bed. The fossilifer-
ous strata represent turbidites interbedded with dark gray
mudstone and siltstone. Cones and sticks were buried rapidly
and preserved without alteration. Some of the broken and
incomplete cones were evidently partly decomposed prior to
displacement into the marine basin. Most of the cones have
suffered compression, owing to the thick sediment load that
accumulated above them, as well as to intense folding and
thrusting of the section (Bailey and Jahns, 1954, fig. 8;
Jennings and Troxel, 1954, fig. 17; Yeats et al., 1981, figs.
1-3).

As suggested for the Drakes Bay site in central California (Axelrod, 1980a, pp. 21-23) and a comparable deposit 32
km east in the Ventura Basin near Santa Paula (Crowell, 1967,
pp. 999-1000), the cones and branches probably accumulated in
a coastal lagoon at the mouth of a major river. The debris
was then swept seaward in a turbidity current that may have
originated from exceptional flooding during a major storm.
It is also possible that the turbidite was triggered by an
earthquake, for there are numerous large faults in the area
(Bailey and Jahns, 1954, p. 59; Jennings and Strand, 1969).
Another possibility is that the deposit was emplaced by an
earthquake-induced tsunami, as suggested for deposits of
small coconuts in turbidites of the Miocene marine section,
North Island, New Zealand (Ballance et al., 1981). In any
event, the debris accumulated in deep water, well off-shore,
where the turbidite came to rest.

Samples of the plant-bearing strata were soaked in water
to disintegrate the compacted rock so that it could be
examined for seeds, but none were found. This is not surprising, for they were probably transported farther offshore,
the lightweight seeds bypassing the heavier sticks and
cones.

According to Robert L. Yerkes and Andre Sarna-Wojcicki
of the U.S. Geological Survey, Menlo Park, who have recently
completed geologic mapping in this area, the rocks that yield
the fossils are in the upper Pico Formation and are probably
a correlative of the Santa Barbara Formation in its type area
20 km west (Dibblee, 1966); the latter differs lithologically
from the Pico, being much sandier.

During a field conference in October 1980, Sarna-Wojcicki
indicated that the plant beds appear to be stratigraphically
slightly above the "Bailey tuff". This wide-spread, thin
marker bed in the upper Pico Formation has been redated at
1.2 $\pm$ 0.2 m.y. B.P. and occurs near the transition from

Globorotalia tosaensis to G. truncatulinoides (Izett, et al.,
1974); the cone bed is probably in the basal part of the
latter foraminiferal age. According to current evidence, an
age of essentially 1.0 m.y. indicates that the plant bed is
contemporaneous with Nebraskan glaciation of the mid-continent (Evernden and Evernden, 1970, fig. 4; Boellstorff, 1978,
fig. 2). As for its relation to Sierran glaciations, the
deposit is probably contemporaneous with Sherwin Till, underlying Bishop Tuff, which is dated at 0.8 $\pm$ 0.1 m.y. (Izett,
et al., 1974). That the Sherwin is younger than the McGee
Till is apparent from the topographic position and greater
degree of weathering of the latter (Putnam, 1962). The McGee
rests on basalt dated at 2.7 m.y. (Dalrymple, 1963), but the
exact age of the till is not now determinable. The Deadman
Pass tilloid is older, for it lies between lava flows dated
at 2.7 and 3.1 m.y. B.P. (Curry, 1966, 1971). Huber (1981)
has suggested that the deposit may not be a till, but an old
mudflow breccia (or alluvial fan), presumably related to
uplift of the Ritter Range directly west.

COMPOSITION

The fossil cones that make up the Seacliff assemblage are
inseparable from those produced by living species, as is true
for most woody taxa whose records range well down into the
Miocene. The relationships of the Seacliff species to previously described Tertiary species are as follows:

Seacliff Species	Tertiary Species
Pinus attenuata Lemmon	P. pretuberculata Axelrod
P. muricata var. muricata D. Don	P. masonii Dorf (in part)
P. muricata var. stantonii Axelrod	(not recorded)
P. radiata D. Don	P. lawsoniana Axelrod
P. remorata Mason	(not recorded)
P. sabiniana Douglas	P. pieperi Dorf
Pseudotsuga menziesii (Mirbel) Franco	P. sonomensis Dorf

The relative abundance of cones in the collection is: <u>Pinus</u> <u>muricata</u> var. <u>muricata</u>, 52; <u>P</u>. <u>radiata</u>, 24; <u>Pseudotsuga</u> <u>menziesii</u>, 6; <u>Pinus</u> <u>muricata</u> var. <u>stantonii</u>, 4; <u>P</u>. <u>remorata</u>, 4; <u>P</u>. <u>attenuata</u>, 4; <u>P</u>. <u>sabiniana</u>, 2. Inasmuch as they were transported into the marine basin from different sites on the adjacent mainland to the north, these cones provide us with information regarding the regional distribution of vegetation and climate in western Ventura Basin.

The occurrence of six pine species in the Seacliff flora points up the limitations of palynology in Quaternary and older rocks. Heusser (1978) recorded abundant pine pollen in a late Pleistocene core recovered from the Santa Barbara Basin, 55 km west-southwest of the Seacliff site, but no species was identified. Actually, there are seven pine species in the drainage basins that empty into Santa Barbara Basin today and which may contribute pollen to accumulating sediments. Since the pines range from sea level to above 1,800 m, recognition of <u>Pinus</u> pollen in a core from Santa Barbara Basin (or elsewhere) obviously provides only meager floristic information and little or no ecologic or climatic insight. It is regrettable that palynologists have not applied the method developed by Ting (1966) to identify pine pollen. Ting concentrated his effort on discriminating between the 20-odd species of California pines and was quite successful. Several skeptics forwarded him a total of 50 modern pine samples and 48 were identified correctly. Of the two that were not identified, one was a hybrid in which one of the parents was recognized and the other was a species whose identity was not positively known to the collector.

PALEOECOLOGY

Vegetation

Judging from the ecologic requirements of the species, the conifers represent three dissimilar ecologic groups--coastal, interior, and upland. Cones of species of present coastal occurrence are most abundant in the Seacliff collection, with Pinus muricata represented by 52 specimens and P. radiata by 24. This suggests that these species formed dense forests on the seaward slopes bordering the shore, where precipitation was adequate and fog persistent in summer. From such an area, their cones would be contributed in relatively larger numbers to a lagoonal site. Pseudotsuga menziesii, represented by 6 cones, probably occurred in sites with higher rainfall on the coastward slope, above and inland from closed-cone pine forest, much as can be seen today in central California, where the forests are ecotonal (Pl. 4, fig. 2). Pinus remorata is represented by 4 cones, consistent with its evident preference for well-drained, drier sites, as shown by present populations near San Luis Obispo (Beacon 859), Lompoc (Pine Canyon), on Santa Cruz and Santa Rosa islands, on altered volcanic rocks in entrenched meanders of the coastal terrace 5 km south of Eréndira, Baja California, and near the summit of Cerro Las Piñatas (elev. 550 m) on a soilless rocky substrate. P. muricata var. stantonii, represented by 4 cones, may have lived in drier sites bordering the shore, as inferred from its present occurrence in the lee of Santa Cruz Island.

The species that now occur in upland (Pinus attenuata) or inland (P. sabiniana) areas are represented by 4 and 2 specimens, respectively. Their low frequency may be attributed to transport from more distant sites in hills bordering the Pico embayment. P. attenuata presumably grew on the forest border in sites with thin soil on drier, southerly slopes with greater ranges of temperature, as is typical for the species today. P. sabiniana was a member of a pine-oak woodland-grassland, as judged from its present occurrences in central and northern California and from its fossil associates in the late Tertiary Mount Eden (Axelrod, 1937; 1950b) and Anaverde (Axelrod, 1950c) floras of southern California. It probably inhabited drier sites on warmer slopes bordering Ventura Basin to the north and east, becoming more abundant inland as ranges of temperature increased and precipitation decreased.

This assemblage finds a general parallel in the distribution of vegetation in the vicinity of Año Nuevo Point, north of Santa Cruz. Pinus radiata forms a dense forest along the coast, reaching inland 1 or 2 km, where it interfingers with the lower margin of the Pseudotsuga-Sequoia forest; no Sequoia cones were recovered from the Seacliff deposit. This ecotone is well exposed in the valley of Waddell Creek (Pl. 4, fig. 2) and also in the valley of Scott Creek along old Highway 1 near Swanton. Dry, south- and west-facing slopes of hard, cherty Monterey shale, directly above the valley in which P. radiata and Pseudotsuga occur, support stands of Pinus attenuata, occasional trees of which are associated with these conifers.

Pinus muricata is absent from the Año Nuevo area but occurs 120 km north on Inverness Ridge, where it is associated with Pseudotsuga, and 60 km south on Huckleberry Hill at Monterey, with Pinus radiata; Pseudotsuga is found farther inland and south in the Santa Lucia Range. Pinus sabiniana occurs inland in the Santa Cruz Mountains, in areas with greater ranges of temperature, as well as at many localities throughout the inner Coast Ranges, reaching south to the margin of southern California in the Santa Ynez Valley north of Santa Barbara.

Although the areas around Santa Cruz and Monterey seem generally to represent the flora that lived on the coastal slope of the Ventura Basin north of Seacliff, there are important differences. The P. radiata populations at Año Nuevo and Monterey have larger cones than those from Seacliff, which are the mean size of the Guadalupe Island population, var. binata. Furthermore, cones of P. attenuata in the Santa Cruz Mountains are larger and the apophyses are not so attenuated or flattened as in the Seacliff specimens, which are more like those produced by the var. acuta in the mountains of southern California. Stands of P. remorata are known only from local sites in coastal and insular southern California and Baja California, where it is associated with P. muricata.

The evidence thus suggests that we are dealing with an overlap of members of different floras. Pseudotsuga menziesii apparently ranged southward along the coast to overlap the coastal Pinus muricata, P. radiata, and P. remorata. This agrees with the occurrence of Pseudotsuga from central California northward during the Pliocene and earlier, and with the presence of the pines in southern California in the early Pleistocene and earlier. Cones of P. radiata at Veronica Springs Quarry, Santa Barbara, are the size of those at Seacliff, and both are within the mean size of the Guadalupe Island population (Axelrod, 1980a). There is a similar species in the Mount Eden flora (Axelrod, 1937), dated at about 5 m.y. (Matti and Morton, 1975). P. remorata occurs in Plio-Pleistocene transition beds at Potrero Canyon, Santa Monica (Axelrod, 1967), and at Century City in west Los Angeles (Axelrod, 1980a). P. attenuata is represented by var. acuta (as P. linguiformis Mason) at Signal Hill in Long Beach (Mason, 1932), also of transitional Plio-Pleistocene age. The allied P. pretuberculata is in the Mount Eden flora (Axelrod, 1937), which also contains fossil digger pine (P. pieperii Dorf). The latter species has also been recorded from the Pliocene lower Pico Formation and from Plio-Pleistocene rocks in Lake Canyon near Ventura (Wiggins, 1951). Its occurrence at these latter two sites may be attributed to transport from warmer and drier interior areas.

That P. muricata ranged south with Pseudotsuga seems unlikely. Its oldest records are in the lower Pico Formation as P. masonii Dorf and in the middle Merced Formation of coast-central California (Axelrod, 1980a), records that are Pliocene in age. When it reached central California is not now known. It may have spread northward along the coastal strip during the warm late Miocene (mid-Hemphillian-Messinian) as a maritime member of the Madrean flora that is well represented in the interior, as in the Mulholland,

Oakdale, and other floras (reviewed in Axelrod, 1977; 1980b). This is consistent with its occurrence in the middle Merced Formation just north of the San Andreas rift, in rocks about three million years old. This species thus appears to have been distributed widely along the coastal strip during the late Miocene. Its probable later history is considered below (see Evolution).

Climate

That climate was suited to the southward shift of Pseudo-tsuga menziesii may be inferred from the nature of the early Pleistocene Soboba flora at the northwestern base of the San Jacinto Mountains, 235 km east-southeast of Seacliff in interior southern California (Axelrod, 1966). Megafossil remains of typical mixed conifer forest trees, notably Abies concolor (Gord. & Glend.) Lindley, Calocedrus decurrens (Torrey) Florin, Pinus lambertiana Douglas, P. ponderosa Lawson, and Populus tremuloides Michaux, occur in this flora. This suggests that the forest belt was about 1,000 m lower than that now in the adjacent mountains. It also indicates that precipitation was 350-500 mm higher than that in the Soboba area today (300 mm), and that mean annual temperature was about 5.5°C lower (Axelrod, 1976, fig. 5). Such condi-tions would have been suited then to the southward shift of Pseudotsuga menziesii along the coastal strip. The Soboba flora occurs in the Bautista Formation, dated as Irvingtonian by its mammalian fauna (Savage, 1951). It is therefore essentially contemporaneous with the Seacliff flora and with Nebraskan glaciation, as judged from recent evidence of its age discussed in Chapter I.

To support the mixture of conifers at Seacliff, precipi-tation must have been higher than at present. A minimum of 760 mm (30 in) along the coast may be inferred for Pseudo-tsuga. Rainfall no doubt increased up the coastal slope, where it may have reached 1000 mm (40 in) or more. Since this flora evidently lived during a pluvial contemporaneous with the Nebraskan glaciation, there was probably heavy fog during summer. At present, rainfall in the Seacliff area is 350 mm, as judged from coastal meteorological stations at Oxnard and Santa Barbara, 38 and 20 km to the east and west, respectively. This implies that rainfall at sea level has been halved, the decrease amounting to fully 350-400 mm (13-15 in) annually. Whereas forest formerly covered the Pico shore and bordering slopes, today the area is covered with a dense Venturan Sage scrub (Axelrod, 1978, fig. 2; Pl. 4, fig. 1 of this report). The only trees in the nearby area are riparian species of Juglans californica S. Watson, Platanus racemosa Nuttall, Populus fremontii S. Watson, and Salix lasiandra Bentham along major river courses and, with an occasional live oak (Quercus agrifolia Nee) reaching to near the coast on sheltered north slopes in river valleys.

Pseudotsuga menziesii indicates that temperature was lower than that now in this part of coastal southern California. Mean temperature on the coastal slope may be inferred from that at Santa Cruz and Monterey to the north, as compared with Santa Barbara and Oxnard (Table 4). Charting the

TABLE 4

Mean temperature (°C) data for sites in coastal northern and southern California

STATIONS		MEAN ANNUAL	WARMEST MONTH	COLDEST MONTH
Northern California				
Santa Cruz		13.8	17.4	9.4
Monterey		13.5	16.8	9.8
	Mean	13.6	17.1	9.6
Southern California				
Santa Barbara		14.9	18.4	10.8
Oxnard		15.2	18.3	11.8
	Mean	15.0	18.3	11.3

data on a Bailey monogram (Fig. 6) shows that mean warmth of climate has increased from about W 13.7°C (174 days with mean temperature warmer than 13.7°C) to W 14.7°C (203 days warmer than 14.7°C) [Bailey, 1960, 1964]. There has been little change in equability, for all stations have a rating of M 70 or higher. Mean annual temperature was approximately 1.5°C lower than at present (Fig. 6). This is about 2.5°C less than that implied by the generally contemporaneous Soboba flora from interior southern California (Axelrod 1966, 1976, fig. 5). The difference is expectable, since the highly equable climate (M 70+) on the coast would reduce the amount of temperature change. Furthermore, a greater difference

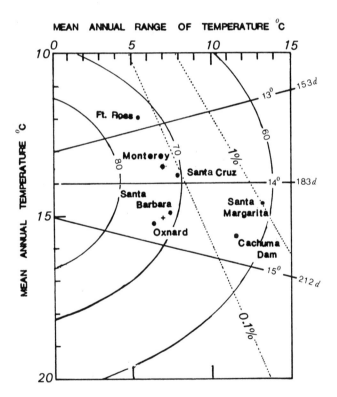

FIGURE 6. Comparison of thermal conditions at Santa Barbara and Oxnard near the Seacliff locality (+) with those at Santa Cruz and Monterey 400 km north, where four coastal conifers now occur that are recorded in the Seacliff flora.

inland has resulted from the later uplift of the Santa Ana
Mountains to the west, which isolated San Jacinto Valley from
the coast. This brought warmer climate to the lowlands and a
cooler one to the forest in the mountains, the difference
amounting to about 1.0°C more than that which would otherwise
be there. This is evident from the higher mean annual temperature in the montane conifer forest on the coastward
slopes of southern California mountains (as at Nellie, Mt.
Wilson) than at stations in more interior locations (as at
Lake Arrowhead, Idyllwild), as shown in Table 5. The difference in warmth between these stations is W 11.0°C (84 days
warmer than) for the interior sites, as compared with W
12.3°C (132 days warmer than) for stations on coastward
slopes open to some marine influence.

TABLE 5

Comparison of temperatures (°C) in lower part of
mixed conifer forest, southern California

METEOROLOGICAL STATIONS		MEAN TEMPERATURE		
		Annual	January	July
Coastward slopes				
Mt. Wilson (1740 m)		12.8	5.4	22.2
Nellie (1525 m)		11.8	3.8	21.4
	Mean	12.3	4.6	21.8
Interior				
Idyllwild (1645 m)		10.9	4.4	19.8
Lake Arrowhead (1586 m)		11.1	2.9	20.6
	Mean	11.0	3.6	20.2

Pinus sabiniana occurs chiefly away from the coast,
inhabiting areas with a greater range of temperature. The
nearest meteorological station in a P. sabiniana community
is at Cachuma Lake (elev. 238 m), 55 km west-northwest of
Seacliff in the Santa Ynez Valley north of Santa Barbara.
Precipitation there totals 450 mm annually but increases with
elevation, as on Figueroa Mountain, 10 km north, where P.

sabiniana and its associates range up to meet the P. coulteri
D. Don-Pseudotsuga macrocarpa (Vasey) Mayr forest in an area
with well over 800 mm precipitation. Similar relations exist
at Santa Margarita, 160 km north at 380 m elevation, where
precipitation totals 835 mm. Mean annual temperature at
Cachuma Lake is 15.6°C and mean annual range of temperature
is 11.7°C. These are 0.5°C and 6°C higher, respectively,
than at the fossil site today, and greater differences are
implied for the P. sabiniana association at higher elevations, as on Figueroa Mountain or to the north in the inner
Coast Ranges. Similar changes may be inferred for areas well
inland from the Pico shore where P. sabiniana no doubt lived.

EVOLUTION

Taxa

The present collection provides additional information
regarding the evolution of closed-cone pines (Pinus subsect.
Oocarpeae), and adds also to an understanding of the history
of the taxa associated with them.

Pinus attenuata

The present record establishes this pine in the Ventura
Basin, situated between its present occurrences to the north
and south in central and southern California (see Griffin and
Critchfield, 1972, map 42). It is allied to P. pretuberculata Axelrod, recorded in rocks 11 m.y. old in the Table
Mountain flora near Sonora in the central Sierra Nevada
(Condit, 1944; Noble et al., 1974). Other records of P.
pretuberculata include the Mount Eden flora near Beaumont,
interior southern California (Axelrod, 1937), dated at 5
m.y.; the Verdi flora west of Reno (Axelrod, 1958), dated at
5.7 m.y. (Evernden and James, 1964); and in Plio-Pleistocene
rocks at Signal Hill, Long Beach, where it was described as
P. linguiformis (Mason, 1932).

As noted below (see Systematic Descriptions), the present
specimens have attenuated, flattened apophyses that distin-
guish the southern California populations in the Santa Ana
and San Bernardino mountains that make up the variety acuta
Mayr (1890). It is not known if seeds of fossil attenuata
were smaller than those of the modern, which might give the
latter a greater selective advantage under a trend to increased drought. The near-coastal fossil occurrences of
cones like P. attenuata var. acuta may represent a relict,
ancestral population. On this basis, the living attenuata
may have originated from P. pretuberculata earlier in the
interior, which is consistent with its presence in the
Table Mountain, Mount Eden, and Verdi floras, all in areas
where climate was less equable than on the coast. Clearly,
future collections are required to test the validity of this
hypothesis.

Pinus muricata

The discontinuous living populations of P. muricata were
described first by Duffield (1951), who concluded that the
species has four varieties. 1. The var. borealis on the
north coast forms large robust trees--the foliage is dense
and dark bluish-green, the bark highly fissured and dark
blackish, the asymmetrical cones reflexed and generally
ovate, with upswept swollen (conical) apophyses, and the
population shows less variation than the other varieties. 2.
The var. typica of Duffield is of moderate size--the foliage
is lighter green and not so dense, the fissured bark lighter
and often grayish, and the cones are variable, ranging from
ovate, asymmetrical, and reflexed with prominent apophyses
to symmetrical, elliptic, and with thin plane scales without
apophyses; although not discussed by Duffield, the latter
cannot be separated from cones in his var. remorata. 3. The
var. remorata is of moderate size--the bark is light gray to
light brown, lightly fissured, the cones symmetrical and
elliptic-ovate to long-conical, lacking apophyses and with

prickles, and attached essentially at right angles to the stem; and the smaller branches are more brittle than those of the other varieties.   4. The var. cedrosensis Howell has consistently rounded, swollen apophyses with minute or no prickles; this is a variety of radiata, not muricata.

In their comprehensive study of the pine populations from Inverness southward, Linhart et al. (1967) also concluded that muricata displays continuous variation from muricata to remorata; they did not include var. borealis or var. cedrosensis in their study.

My work on the fossil pines of the muricata complex raises a major question about these interpretations.   Nowhere does the fossil record show a range of cone variation like that depicted by Duffield or by Linhart et al., one involving a complete range of cone type from remorata to muricata.   The absence from the known fossil record of cones showing such variation, even at fossil sites (Carpinteria, Point Sal) near where muricata lives today and where both remorata and muricata fossil cones have been recovered, suggests that the variation in muricata illustrated by Duffield and Linhart et al. may be the result of postglacial hybridization-introgression.

Renewed interest in this problem was raised by discovery of the Seacliff flora, because abundant cones similar to those of muricata occur there.   In addition, rare cones of remorata are also in the Seacliff deposit, but the variation depicted by Duffield and by Linhart et al. for muricata is not present in this large collection.   Since it represents an accumulation of cones transported initially by river to the coastal site prior to displacement into the marine Pico Basin, the cones are presumably representative of the forest population from which they were derived.   The same argument applies to the rich muricata samples at Carpinteria and Point Sal.   In order to appraise the affinities of the fossil populations more effectively, I sampled the living populations of

muricata from Fort Bragg southward to near Eréndira, Baja
California, and also on Santa Cruz Island, during the period
of mid-December 1981 to early March 1982. The results
of that study differ in some ways from my earlier report
(Axelrod, 1980a), which was influenced by the opinions of
Duffield and Linhart et al. My present view has been shaped
·also by the presumed nature of the type specimen, which comes
from the vicinity of San Luis Obispo (illustrated by Lambert,
1828-37 ed., vol. 3, pl. 55; also 1837-42 ed., vol. 3, pl.
80; a copy from the 1837 ed. appears here as Pl. 5). The
type specimen is misplaced or lost, according to letters of
inquiry made of the herbaria at the Linnean Society of
London, the British Museum (Natural History), and the Royal
Botanic Gardens, Kew. Certainly, D. Don (1836) did not des-
cribe any variation in the cone material Coulter sent him,
and Shaw (1914) also accepted only limited variation. P.
muricata, as I recognize it, has the properties originally
assigned to it: cones ovate-elliptic, strongly asymmetric
with prominent triangular apophyses. Here is Don's description:

2. PINUS MURICATA

P. foliis ternis strobilis inaequilateri-ovatis
   aggregatis: squamis cuneatis apice dilatatis
   umbilico elevato mucronatis; baseos externae
   elongatis ancipiti-compressis recurvato-
   patentibus.

Habitat in California ad locum San Luis Obispo
   Hispanice dictum, alt. 3000 ped. Coulter.
   (v.s. sp.)

Arbor recta, mediocris, altitudine circiter 40 pedes.
   Strobili aggregati (2.v.s.), inaequilateri-ovati,
   3-pollicares: squamis cuneatis, crassissimis,
   apice dilatatis, obsolete 4-angularibus, umbilico
   elevato mucronatis; baseos externae elongatis,
   ancipiti-compressis, callosis, rigidis, laevibus,
   nitidis, recurvato-patentibus.

Trees that produce cones of this sort are abundant in the
southern groves, and those that display variation toward
remorata-type cones are also frequent.

The results of my present study can be summarized briefly in terms of general conclusions.

1. Present <u>muricata</u> populations show that a range of cone variation from one extreme (var. <u>muricata</u>) to the other (<u>remorata</u>) are the result of hybridization-introgression. In proceeding northward from southern California to the Inverness and Fort Ross groves, the evidence of <u>remorata</u> there is a few genes (presumably) that control cone shape. To consider this hybridization, we would expect both parents present with their more-or-less intermediate hybrid offspring. Although this does occur in the populations from San Luis Obispo southward, there are few parental <u>remorata</u> trees (based on terpene, resin canal, enzyme data) in the northern populations, just their back-, back-, back-cross introgressive progeny. That is, the northern populations apparently represent segregation products of hybridization farther south. Although Mason (1949) felt that <u>remorata</u> is being submerged by <u>muricata</u>, Connie Millar points out (March 1982) that it is doing better than earlier, since hybridization-introgressive relations indicate that its influence is spreading, or has recently done so. Recall that fossil records of <u>remorata</u> occur on Santa Cruz Island, Point Sal, Carpinteria, Seacliff, Potrero Canyon, and Century City, all centered around south-central to insular southern California. None show any evidence of gradations toward var. <u>muricata</u>. At present, pure <u>remorata</u> populations range from Baja California (Eréndira) to Purisima Hills and near San Luis Obispo. Only occasional trees with typical <u>remorata</u> cones occur as far north as Fort Ross. This northward spread presumably occurred during the Xerothermic which favored the northward and coastward shift of taxa now typical of warmer, more southern climes.

2. The var. <u>borealis</u> is a distinct taxon with the features noted above. Of the several populations measured by Duffield, it is the least variable in its characters. At

present, its principal southern area is Inverness Ridge,
Marin County, where the trees are dark green and well foli-
ated, the bark dark and deeply furrowed, and the cones have
the distinctive swollen (conical) upswept apophyses. There
are also scattered trees of this variety along the crest of
San Geronimo Ridge, south of Lagunitas, Marin County. The
somewhat smaller stature of these trees, as compared with
those of the coast farther north, is probably due to the
lower precipitation and lower fog frequency. The southern
most site at which a few trees referable to var. borealis
occur is Huckleberry Hill, Monterey, as noted in Chapter I.

    3. The var. stantonii (a new taxon) is confined to Santa
Cruz Island. The principal populations occur above Chinese
Harbor on Monterey shale, on the shale on the upper south
slope of the ridge above Chinese Harbor, and on basalt and
andesite above Pelican Bay. At all these sites it is associated with P. remorata and is interbreeding with it. Var.
stantonii differs from muricata var. muricata in having fewer
cone scales, the apophyses are broadly deltoid distally and
often are directed outward, and the cones have a more massive
appearance, in spite of their moderate dimensions (Pls. 6 and
7). The relatively larger cones of the insular population
parallel a similar trend to increased size in many insular
taxa that have close mainland relatives, whether herbs or
shrubs.

    4. P. remorata is now known to form essentially pure
stands at a number of sites. Among these are Santa Rosa
Island; western Santa Cruz Island; Coon Creek south of Morro
Bay, a population difficult to sample because of the steep
slopes and the dense stand with tall trees, the latter with
light gray, relatively smooth bark; Hilltop 859, 6.5 km (4
mi) south of San Luis Obispo and 2 km (1.5 mi) east of Highway 101; on top of the terrace directly west and southwest
of the gate to Vandenburg Airbase in Pine Canyon, northwest
of Lompoc; in Cañon de los Pinitos, about 0.5 to 1.5 km from

the coast, in the entrenched second coastal terrace at an
elevation of 60-90 m (200-300 ft), 6.5 km (4 mi) south of
Eréndira; and on the upper north slopes of Cerro Las Pinitas,
at an elevation of 500-600 m (1,600-1,800 ft), about 9.5 km
(6 mi) southeast of Eréndira. Cones indistinguishable from
those of P. remorata occur in two deposits of transitional
Plio-Pleistocene age in southern California, at Potrero
Canyon in Santa Monica and Century City in West Los Angeles,
as well as at several late Pleistocene sites.

 Discussion. With respect to evolution of P. muricata and
its varieties, P. masonii Dorf (1930) provides us with some
information. The type material comes from the lower Pico
Formation west of Ventura in rocks about 3-4 m.y. old; one of
the cotypes (no. 307, Pl. 5, fig. 6) is not allied to muricata, but to radiata. P. masonii also occurs at two sites in
the Merced Formation on the outer coast south of San Francisco in rocks dated at 4-2 m.y. old (Dorf, 1930; Glen, 1959;
Axelrod, 1980a). These records are known from only one or
two cones and do not give us a good idea of the variation of
the species. Nonetheless, it is apparent that a species pro-
ducing cones similar to those of var. muricata was widely
distributed in the coastal strip. These cones are not all
exactly the same but show moderate differences in size, shape
and development of triangular apophyses.

 In view of these differences, as well as its wide distribution, it is inferred that masonii (or muricata) was ancestral to the two modern varieties, borealis and stantonii. On
the basis of its distinct morphology, borealis may have
separated earlier from an ancestral population. P. muricata
var. muricata now occupies discontinuous areas along the
central coast, reaching southward into Baja California, where
it was described as the var. anthonyi (Lemmon, 1895). The
type specimen (U.C. Herbarium) differs in no important way
from var. muricata, as described originally (Don, 1836) and
illustrated by Lambert (1837; see Pl. 5).

As noted earlier by Mason (1949, p. 360), and which I have also observed on several occasions, the remorata influence in modern muricata populations increases southward from Monterey, though it is also present in Marin County. This not only involves cone morphology, but is evident also in the more open, less luxuriant, lighter green foliage, in the stature of the trees, and in the change from very dark, deeply furrowed bark to lighter gray-brown or gray bark with shallower fissures. The evidence suggests that var. borealis is now at the north end of an ancient cline (i.e., masonii Dorf) that ranged from coastal north-central California southward into Baja California during the late Tertiary. By the Quaternary, masonii had given way imperceptibly to P. muricata var. muricata, which presumably ranged southward into Baja California. A segregate (var. stantonii) with larger, broadly deltoid apophyses and fewer cone scales has persisted on Santa Cruz Island. Judging from its Seacliff occurrence, it appears to represent an older continental species that was restricted to the island like other insular endemics (i.e. Ceanothus arboreus Green, Lyonothamnus asplenifolius Greene, Prunus lyonii (East.) Sargent) that have been recorded on the continent.

Finally, it is reiteriated that P. remorata has been a distinct species since at least the Pliocene-Pleistocene transition, and is certainly considerably older. It now crosses with P. muricata and the vars. borealis and stantonii (Fig. 7), but this is judged to be a postglacial event because the range of variation displayed by the hybrids has not been recovered from the fossil record. The reason for the lack of widespread variation in cone type in the fossil samples may lie in environmental change. Some 35 genera of large mammals (horse, mastodon, ground sloth, llama, glyptodon, camel, sabretooth cat, etc.) became extinct in the United States (Hibbard, et al., 1965; also Martin and Guilday, 1967), as did a number of birds. This coincided

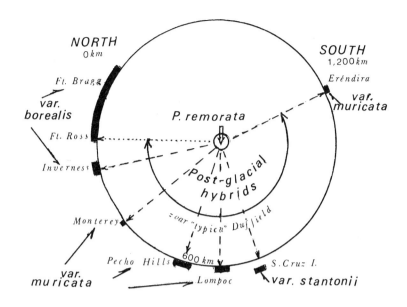

FIGURE 7. Inferred interrelations of <u>Pinus</u> <u>muricata</u> and its varieties (see Plates 6 and 7). Times of separation of taxa are not known.

with the retreat of pinon-juniper woodlands from the present desert and semidesert areas, the wide restriction of grass-lands, the rapid spread of regional desert environments, and the emergence of a more pronounced summer dry season as the intensity of the mediterranean-type climate increased. At the same time, numerous lakes disappeared, springs dried up, water-courses became intermittent over wide areas. The resulting faunal extinction was greater than that at any other time in the Quaternary. By contrast, the plants did not undergo extinction; the principal changes were range dis-junctions. Taxa of mesic to subhumid requirements were con-fined to relict areas in the interior or to the coastal region (see below). Increasing extremes of temperature and decreasing precipitation enabled desert and interior taxa to spread to the coastal strip in southern and central Califor-nia, respectively. Inasmuch as <u>P. remorata</u> is adapted to relatively drier sites, the emergence of a more extreme medi-terranean-type climate may have enabled remorata to freely interbreed with populations of P. muricta var. muricata, borealis, and stantonii. Certainly, if any variation in these taxa occurred earlier, it must have been confined to very local sites.

The recent emergence of hybrid swarms from Inverness southward may reflect their greater adaptability to a more severe postglacial mediterranean-type climate, a relation consistent with the preference of remorata for warmer, drier climates. The var. borealis in the far north (Fort Bragg-Sea Ranch) has been isolated sufficiently so that it not only is "completely isolated by reproductive barriers from the southern elements of the species complex" (Critchfield, 1967, p. 94), it differs in structure, foliage color ("blue" population), and bark as well. Nonetheless, it is significant that there is a gradual change in the borealis population southward, as shown by occasional trees with more prominent apophyses in the Fort Ross area and by its ready hybridization with remorata on Inverness Ridge, where the cones are more prominently armed and more elliptic than those to the north. The influence of remorata decreases northward, presumably because the cooler and moister climate there provides fewer of the warmer, drier sites that remorata appears to favor.

My conclusions with respect to muricata, based on largely different evidence, agree with those expressed earlier by Duffield (1951, p. 53):

> Perhaps there were pre-Pleistocene hybridizations with subsequent divergences, yet it is not necessary to postulate (them) and there appears to be no compelling evidence that they occurred. It is possible that today we are witnessing the first natural hybridization in this group....Pinus muricata, moreover, appears to be in the process of diverging into several recognizable entities...

The recency of hybridization in the muricata complex apparently finds a parallel in various taiga conifers of

Eurasia that display introgressive hybridization of the domi-
nants (Bobrov, 1983). The hybrids of Picea obovata and P.
abies (=Picea X fennica), Larix kamtschatica and L. gmelinii
(= Larix X maritima), Pinus densiflora and P. sylvestris
(= Pinus X funebris) and others reflect substitution in
environmentally intermediate or newly available areas.
Bobrov considers that the changes occurred at the end of the
Pleistocene, and especially in the Holocene.

Pinus radiata

The 24 cones of P. radiata are the same mean size as
cones of the present Guadalupe Island population, P. radiata
var. binata (Pl. 9; also Axelrod, 1980a, pl. 1 and fig. 2).
Together with those of similar size recorded previously from
the early Pleistocene at Veronica Springs Quarry, Santa
Barbara, they support the inference that there has been a
trend to larger cones in the radiata line of evolution.
Significant in this connection is the occurrence of abundant
cones of P. radiata at Carpinteria (Chaney and Mason, 1933),
10 km west. The Carpinteria cones, much younger in age
(40,000 B.P.), average considerably larger (10.9 vs. 8.0 cm
long) than those from Seacliff (Fig. 8). With respect
to the other populations, Figure 8 shows that they differ
importantly in mean cone size, with those at Año Nuevo and
Cambria displaced farther to the right, to larger cones than
those of the Monterey and Guadalupe Island populations. The
progressive increase in mean cone size is:

    Guadalupe I. (8.2 cm.) -- Monterey (9.6 cm.) = 1.4 cm.
    Monterey (9.6 cm.) -- Año Nuevo (11.6 cm.) = 1.8 cm.
    Ano Nuevo (11.6 cm.) -- Cambria (13.9 cm.) = 2.5 cm.

That the size differential may even be somewhat greater for
the Cambria population is implied by a mean size of 14.3 cm
reported by Fielding (1953), as compared with 13.9 cm for my
sample of 352 cones. The relations suggest greater stability
for the Monterey and Guadalupe Island populations and strong
selection for larger cones. With their larger seeds, the Año

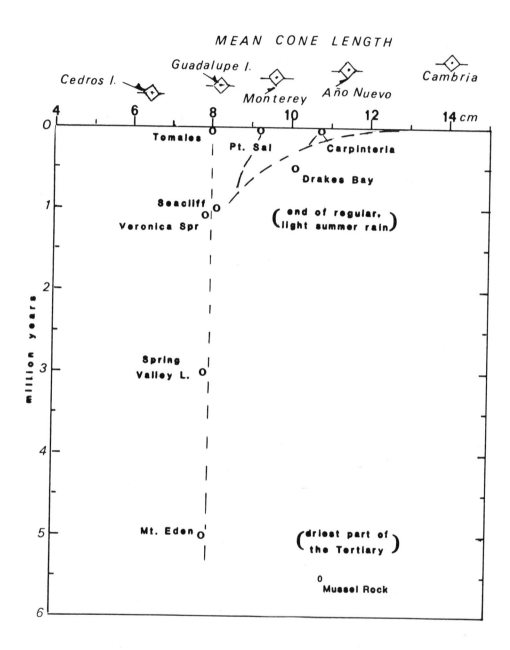

FIGURE 8. Comparison of mean cone length of modern and fossil populations of <u>Pinus</u> <u>radiata</u> (from Axelrod, 1980a, Fig. 2). When charted against time, increase in cone size appears to have accelerated during the latest Pleistocene.

Nuevo and Cambria populations would have greater survival value (see Baker, 1972), especially under the influence of increasing summer drought.

The question arises as to when the trend to larger cones commenced. Cones the size of the Cedros Island population have not been recovered as fossils. However, cones the mean size of the Guadalupe Island pines are recorded from 5 m.y. (Mt. Eden) up to 29,000 B.P. (Tomales flora), as charted in Figure 8. Cones the size of the Monterey population occur at Drakes Bay, now estimated to be 0.5 m.y. old, and at Point Sal dated at 28,000 B.P. Cones from the rich Carpinteria deposit are intermediate between the Monterey and Año Nuevo populations. Cones the mean size of the Cambria population have not been recovered as fossil, though cones of other radiata populations are known from a number of sites along the coast. The apparent absence of Cambria-size cones in the record implies that that population may be of more recent origin. The data are admittedly incomplete and meager, yet they seem consistent with the suggestion that cone size increased rapidly in the late Pleistocene. By that time, light summer rain had been eliminated and mediterranean-type climate had become more severe, with its increasingly longer period of summer drought. This was intensified by the strengthening of the Tonopah High that brings hot, dry Santa Ana winds to the coastal strip in late summer and autumn, crucial times for seedling survival. Since cones of the Cambria population are considerably larger than those of the Año Nuevo, they may reflect an accelerated postglacial response to much drier summers in that area. As noted elsewhere (Axelrod, 1982), the Cambria population is in a drier area than either the Monterey or Año Nuevo groves, and a similar relationship no doubt existed during the Xerothermic. At that time, greater drought stress in the Cambria area may have stimulated a more rapid response to increased cone and seed size than in populations to the north, which were (and are) under more humid climates. The smaller size of the Monterey population seemingly reflects its position at the head of the Monterey submarine canyon which provides greater fog frequency there, and hence has sheltered a more ancient, ancestral population (Axelrod, 1982). By contrast, the Año Nuevo population to the north was also undergoing increased cone size during the Xerothermic, but rate of increase was not so great as to the south, where Cambria was under a warmer, drier, more stressful climate.

A further problem is raised by the record of P. radiata cones at Mussel Rock on the coast south of San Francisco. Cones were recovered there by A. C. Lawson in 1895 from an old forest soil that rests on Franciscan diabase and under-lies the basal beds of the type Merced Formation, which are about 5-6 m.y. old. This occurrence was also noted and discussed by Glen (1959, p. 151). The two cones now available are close to the mean size of the Monterey population (Axelrod, 1967, pl. 1, figs. 1, 2). Their comparatively large size may reflect the warm, dry climate recorded by the wave of Madrean sclerophyll vegetation that spread into central California at this time (Axelrod, 1977, 1980b). This suggests that radiata was sufficiently adaptable to respond to semi-arid conditions earlier in its history. Since climate became moister in the Pliocene (3.5 m.y.), as shown by the Santa Clara (Dorf, 1930) and Sonoma floras (Axelrod, 1944, 1950a), the Mussel Rock population may have ceased to exist. In this regard, it is noteworthy that radiata cones recovered from the younger Santa Clara Formation at Crystal Springs Reservoir (Axelrod, 1967, pl. 3, figs. 3, 4), 18 km southeast, are the mean size of the living Guadalupe population, var. binata. Furthermore, cones of similar size dominate the Millerton flora (Axelrod, 1980a, pls. 2, 3). These relations imply that adjustment of a chronocline to environmental change is necessary for persistence over the long term.

Pinus remorata

There are only 4 cones of Pinus remorata in the Seacliff sample, and the species is also rare at Carpinteria and Point Sal. At all these sites P. muricata and P. radiata are either the common or dominant species. The rarity of remorata cones at these locations may be attributed to its preference for drier, warmer, well-drained sites (Axelrod, 1967, p. 129). Such habitats could be provided by old, elevated marine terraces or rocky slopes with thin soil well removed from the more mesic, dense valley and slope forests of P. radiata and P. muricata. In brief, P. remorata is a distinct and certainly an ancient lineage, consistent with its relative abundance in deposits at Santa Cruz Island (Mason, 1949, p. 359), in Potrero Canyon, Santa Monica, and at Century City, Los Angeles, the latter two of transitional Plio-Pleistocene age. The relatively common remorata cones in the deposit on Willow Creek, Santa Cruz Island, are typical of the species and show no evidence of any muricata influence (i.e., asymmetrical, reflexed cones with strongly developed apophyses). P. remorata dominates the forest patches in the western part of Santa Cruz Island today, and patches occur a short distance upstream from the fossil site. It would appear that the species has persisted unchanged in this area for at least 14,000 years.

P. remorata was initially considered to be an insular endemic confined to Santa Cruz and Santa Rosa islands and was hence termed Island pine (Mason, 1949, p. 358). Essentially pure stands have now been collected at several mainland sites, as noted earlier. As for its occurrence on the Channel Islands, which have been emergent since the later Miocene, this region is very active tectonically. The pine and certain other taxa (including some animals--i.e., salamanders) may have been in the island area when it was united with the mainland (see Azzaroli, 1981), and have since been isolated by tectonic movements.

Pinus sabiniana

The present specimens differ in no significant way from cones produced by the living P. sabiniana, though in size, 12.5 cm (estimated) long and 7.5 cm. wide, one is smaller than the average modern cones. The species has also been recorded from Plio-Pleistocene rocks in Lake Canyon, 8 km (5 mi) east of Ventura (Wiggins, 1951). Older records of the allied P. pieperi Dorf include a record in the lower Pico Formation (Dorf, 1930) and the Mount Eden (Axelrod, 1937) and Anaverde (Axelrod, 1950c) floras of interior southern California, the former now considered about 5 m.y. old (Matti and Morton, 1975). These records show that fossil digger pine was widely distributed in southern California, whereas it now occurs chiefly north of the Transverse Ranges (see Griffin and Critchfield, 1972, map 56). Its present southernmost stations are marginal to southern California, as in Santa Ynez Valley north of Santa Barbara, on Piru Creek above Piru Guard Station (since destroyed by fire), and at the northwest end of Liebre Mountain a few kilometers southeast of Gorman. Its disappearance from sites well within the southern California province may have resulted from the spread of warmer, drier climate during the Xerothermic. This is inferred from the invasion into coastal southern and central California of drier, interior semi-desert taxa at that time, and from the occurrence of the P. sabiniana community in central and northern California in sites now surrounded by mesic conifer forest (Axelrod, 1966, pp. 42-55). The latter stands are presumed relict from the Xerothermic, because they are well isolated from the principal area of the species that now makes up the pine-oak savanna woodland surrounding the Great Valley at lower, warmer, and drier levels (see Axelrod, 1981).

The few fossil cones that have been recovered to date are smaller than the mean of the modern species, which average about 17-18 cm long. If future fossil collections show that

those cones are also smaller than the mean of the modern, it
would indicate that size increased during the later Cenozoic,
a relation shown also by radiata pines. Such an increase in
sabiniana cone size may also be attributed to increasing
summer drought. Although these smaller cones approach those
of P. torreyana in size, there are important differences in
cone shape, apophyses development, and cone-scale cross-
section.

Pseudotsuga menziesii

The Seacliff cones are similar to those of Douglas fir,
a species recorded previously in southern California only
from the Willow Creek flora, western Santa Cruz Island
(Chaney and Mason, 1930), dated at 14,200 years B.P. (Fer-
guson and Libby, 1966). The species probably shifted southward from central California as climate became cooler and
moister, as shown by the nature of the Pliocene Sonoma floras
(Axelrod, 1944, 1950a), dated at 3.5 m.y. Cooler, moister
climate is indicated also by the Santa Clara flora (Dorf,
1930) from near sea level, with montane conifers (Calocedrus,
Pinus cf. lambertiana) as well as Pseudotsuga, dated at 2-3
m.y. (Sarna-Wojeicki, 1976).

The fossil record has not yet clarified the problem of
the origin of Pseudotsuga macrocarpa (Vasey) Mayr, a common
tree of the lower to middle mountain slopes in southern
California. If it was derived from P. sonomensis Dorf (cf.
P. menziesii), this presumably took place earlier in the
Tertiary. One possibility is that the greater cone and seed
size of P. macrocarpa represent a response to drought. If
so, the trend was already completed by the end of the Miocene
(5 m.y.), as shown by cones of P. premacrocarpa Axelrod in
the Mount Eden flora (Axelrod, 1937). Since California west
of the San Andreas fault has moved northward, P. macrocarpa
may have originated from stands of Pseudotsuga further south.
This may have occurred at the present latitude of Baja California, when California west of the San Andreas was a part of

western Mexico--prior to the opening of the Gulf of California. Upland areas were present there, Pseudotsuga was already in the central Rocky Mountains in the late Eocene and Oligocene, and active upland volcanic centers linked that area with western Mexico (McDowell and Clabaugh, 1979, fig. 1; Lipman et al., 1972, fig. 1). There certainly was a protracted dry season over the region by the middle Eocene and during the Oligocene and Miocene as well, as judged from the Green River (MacGinitie, 1969), Florissant (MacGinitie, 1953), and Creede floras (Axelrod, ms) and from diverse evidence regarding the history of Sonoran Desert vegetation (Axelrod, 1979).

In view of this inferred history, it is expectable that P. macrocarpa is confined essentially to the mountains of southern California. Its absence farther north in the Coast Ranges and Tehachapi-Sierra Nevada, which now provide ready avenues for dispersal, is probably the result of colder winters there. This is consistent with its presumed origin in warmer latitudes, and implies that a tolerance for mild winter climate was attained early in its history.

Vegetation

The later Tertiary floras of California, and other regions as well, show that fossil species were not sharply segregated into communities as are their descendant associations (Axelrod, 1977, figs. 5-5, 5-9). As for southern California, fossil species that are allied to taxa of present coastal or insular occurrence reached well into the interior during the late Miocene. This was due in part to marine embayments that moderated climate in the interior as well as to the subdued topography that resulted in a more even distribution of precipitation and temperature. With a more equable climate, taxa intermingled in a broad ecotone that resulted in a mixture of coastal and interior taxa. This is shown by the Mount Eden flora, where Pinus lawsoniana (cf.

radiata) is recorded with other species of present coastal
occurrence, including Ceanothus (cf. spinosus Nuttall),
Juglans (cf. californica S. Watson), Malosma (cf. laurina
Watson) and also Laurocerasus (cf. lyonii [Eastw.] Sargent),
which is now largely insular. . At Mount Eden, the presently
coastal taxa were associated with a Pinus (cf. sabiniana)-
Quercus (cf. douglasii Hooker & Arnott)-Q. (cf. lobata Neé)
woodland-savanna. Desert-border taxa (Cercidium, Condalia,
Dodonaea, Eysenhardtia, Ficus) were on adjacent dry slopes,
and a big-cone forest in the bordering hills was composed of
Pseudotsuga (cf. macrocarpa) and Pinus (cf. coulteri). A
similar intermingling of coastal and interior taxa occurs in
the Piru Gorge flora, with Ceanothus (cf. spinosus) and
Laurocerasus (cf. lyonii) representing the former, and
species of Quercus and diverse chaparral taxa the latter
(Axelrod, 1950d). There is a comparable mixture in the
Anaverde flora (Axelrod, 1950c), with Malosma and Persea
representing taxa that required mild climate. They were
associated with species that contributed to digger pine-live
oak woodland and arid subtropic scrub, including Colubrina,
Dodonaea, and Eysenhardtia.

The intermingling of taxa from chiefly coastal and
interior areas still occurs in southern California. Species
typical of the equable coastal strip, such as Juglans
californica, Malosma laurina, and Rhus integrifolia, occur
where moderating breezes during the warm summers extend up
major river valleys, corridors that also bring milder winter
temperatures to those areas. These taxa extend fully 70 km
up the Santa Clara River drainage to near Castaic, up the San
Antonio River drainage to the vicinity of Altadena, and
similar distances up Brea Canyon to Claremont-Upland. In
addition, Arbutus has relict patches in moister, nearby
upland areas, as in Santa Anita Canyon above Sierra Madre, in
Trabuco Canyon on the coastward slopes of the Santa Ana

Mountains near 1,000 m, and at several sites above avocado
groves on sheltered slopes of the Agua Tibia Mountains east
of Escondido. These relations are paralleled in central
California, where a milder climate from the San Francisco Bay
area extends inland to ameliorate conditions on the west
slope of the lower central Sierra Nevada (see Leighly, 1938;
Bailey, 1964). A number of typical Coast Range species reach
their inland or southern limits there, notably Taxus brevi-
folia Nuttall, Arbutus menziesii Pursh, Baccharis pilularis
ssp. consanguinea (D.C.) Wolf, Castanopsis (Chrysolepis)
chrysophylla (Douglas) A. D.C., Lithocarpus densiflorus (H.&
A.) Rehder, and Quercus agrifolia Neé.

The Neogene floras of southern California included
numerous taxa that required summer rain. Persea is well
represented in the Puente, Modelo, Mount Eden, Anaverde, and
Piru Gorge floras. It not only required summer rain, but
mild, frostless winters. Others with similar requirements
include Colubrina, Dodonaea, Eysenhardtia, Ficus, Juglans,
Populus, Sabal, and Sapindus, which contributed chiefly to
oak-laurel woodland and to subtropic scrub vegetation in
the interior. Under a regime of milder winters and summers
and adequate summer rain, communities of woodland and forest
vegetation were not so distinct as at present. As topographic and accompanying climatic differences increased during the
Quaternary, and as a progressively more severe mediterranean-
type climate developed, new local subclimates came into
existence. It was these novel, modern subclimates that
restricted taxa, segregated the present new communities,
sharpened their boundaries (see Axelrod, 1977, figs. 5-5,
5-9), and provided new subzones for the rapid evolution of
herbaceous perennials and annuals (Axelrod, 1966; Raven and
Axelrod, 1978).

The present collection shows that all the conifers in the
Seacliff assemblage had wider distributions than their modern

descendants. Pinus radiata has cones most like those of the
Guadalupe Island groves; P. muricata var. muricata is now on
the coast, 60 km west and 480 km south; P. muricata var.
stantonii is now restricted to Santa Cruz Island; P. remorata
is confined to relict sites near San Luis Obispo, Lompoc, on
Santa Cruz and Santa Rosa islands, and at two sites near
Eréndira in northern Baja California; Pseudotsuga menziesii
has its principal southern stands on the coastward slopes of
the southern Santa Lucia Mountains well to the north; Pinus
attenuata provides a record of the species in the Ventura
basin, whereas today in southern California it has two relict
stands in the mountains at levels above 1,000 m; and P.
sabiniana has largely been eliminated from the flora of
southern California, occurring only at its northern edge.
Restriction of these trees to their present areas, and
segregation of the modern communities, was an event that
followed the last glaciation, as judged from late Pleistocene
records in southern California.

SUMMARY

The rich fossil cone deposit near Seacliff, coastal
Ventura County, provides the following information with
respect to conifer evolution, ecology, and biogeography
during the early Pleistocene and later.

Pinus attenuata evidently originated in the interior,
whereas a related, possibly ancestral form with thinner
apophyses persisted as a relict in the coastal strip into the
early Pleistocene, retreating into the mountains as lowland
climate became drier. This was accompanied by a trend to
thicker cone scales, more massive cones, and a more distal
development of apophyses leading to the modern var. acuta
Mayr.

P. muricata var. muricata was derived from the Pliocene
P. masonii Dorf, which was widely distributed along the central and southern California coast and presumably reached
into Baja California. Varietal segregates include borealis
on the coast from Inverness north and stantonii on Santa Cruz
Island. The postglacial trend to drier climate evidently
favored the hybridization and introgression of muricata and
its vars., borealis and stantonii, with remorata thus
accounting for much of the variation in the present populations. Their discontinuous distribution reflects the spread
of drier, warmer postglacial climate.

P. radiata cones from Seacliff are the mean size of the
Guadalupe Island var. binata, though the cone scales of the
fossils are somewhat more swollen. Cone size increased during the Quaternary, and the rate apparently accelerated in
the late Quaternary to give rise to the larger-coned Ano
Nuevo and Cambria populations. The Monterey population is
situated at the head of the Monterey submarine canyon, which
provided greater fog frequency than that in the other California groves. It was therefore sheltered from the spreading
Xerothermic climate to which the Año Nuevo and Cambria
populations evidently responded by developing larger cones
and seeds.

P. remorata occurred at several sites in coastal southern
California during the Pliocene-Pleistocene transition and
later. Its rarity in most fossil floras may reflect its
preference for well drained, drier sites away from the mesic
valley and slope coastal forests dominated by P. muricata and
P. radiata. Spreading drought, especially in the Xerothermic,
favored the spread of remorata and its hybridization and
introgression with muricata and its vars. borealis and
stantonii.

Pseudotsuga menziesii contributed chiefly to a forest in
hills above closed-cone pine forest, where precipitation was
higher. It probably ranged into southern California in the
late Pliocene or early Pleistocene as climate became moister

and cooler. Assuming that the large cones and seeds of P.
macrocarpa reflect adaptation to greater aridity, it may have
originated from a menziesii-type ancestor by Oligocene time
in southern California, which was then joined by land to what
is now Sonora, Mexico. Thus the species was in southern
California and Baja California as this area separated from
the Sonoran-Jaliscan mainland and was displaced northward by
movement on the San Andreas and allied rifts during the
Miocene and later.

The distribution of species in California Neogene floras
indicates that the vegetation zones to which they contributed
were mixtures of taxa that formed richer communities than the
present ones. The modern segregated communities emerged in
response to increased topographic diversity, a progressively
longer period of drought, and less equable climates inland,
all of which provided new subzones for novel subtypes of
forests, woodlands, and brushlands. There is considerable
community diversity in these vegetation zones today, but the
diversity of woody taxa is lower because many were eliminated
as summer rain decreased, temperature extremes increased, and
taxa were segregated into new communities of narrower ecologic parameters. However, the new, more open subzones in
the more extreme mediterranean-type climate provided oppor-
tunities for rapid diversification of many annual and herbaceous perennial genera that account for the present richness
of the flora.

SYSTEMATIC DESCRIPTIONS

All specimens are deposited in the Museum
of Paleontology, University of California, Berkeley.

Family PINACEAE

Pinus attenuata Lemmon

(Plate 8, fig. 1)

Pinus attenuata Lemmon. Mining and Sci. Press 64: 45, 1892.

Pinus attenuata Lemmon. Axelrod, in R. N. Philbrick, Proc.
  Symposium on Biology of the California Islands, p. 113,
  pl. 6, fig. 1, 1967 (the type specimen of P. linguiformis
  refigured at natural size).

Pinus linguiformis Mason, Madrono, 2:50, pl. 1, fig. 5, 1932.

   The present record comprises two complete cones and parts
of two others. The complete cones are 9.5-11.2 cm long and
5.0-5.8 cm wide at greatest width. The best-preserved one is
markedly asymmetrical, curved, and with the cone scales on
the outer side having sharply attenuated, somewhat flattened
apophyses. These appear to be broader and more flattened
than those in most populations of P. attenuata, though the
cones of the var. acuta (Mayr, 1890) show relationship. The
chief difference is that the apophyses on the fossil cone
scales are more slender and tongue-shaped and are not so well
developed distally. This implies a trend to cones with
thicker apophyses and their greater development distally.
Inasmuch as typical cones of P. attenuata are recorded in the
interior in the Miocene and later (Condit, 1944; Axelrod,
1937, 1958), P. linguiformis may be a surviving ancestral
population confined to the equable coastal strip.

   In southern California, the var. acuta Mayr occurs in the
Santa Ana Mountains in the Pleasant Peak area and in the San
Bernardino Mountains on the City Creek road above Redlands
(Newcomb, 1962), chiefly at elevations of from 1,000 to
1,200 m. To the north, in central and northern California,
the species descends to lower levels as temperature decreases
and moisture increases.

Fossil records of P. attenuata from the Pleistocene are
at You Bet in the central Sierra Nevada (Mason, 1927) and
Oakland, California (Metcalf, 1923).

Collection. Hypotype no. 6270, homeotype no. 6271.

Pinus muricata D. Don
(Plates 5, 6, and 7,and 8, figs. 2-4)

Pinus muricata D. Don. Linnean Soc. London, Trans. vol. 14,
   p. 441, 1836.

Pinus masonii Dorf., Carnegie Inst. Wash. Pub. 412: 70, pl. 5,
   fig. 4 (upper Merced); fig. 5 (Pico); not fig. 6 which is
   P. radiata; 1930.

Pinus masonii Dorf. Axelrod, In R. N. Philbrick, Proc. Symposium on Biology of California Islands, p. 120, pl. 5,
   fig. 4, 1967. (Santa Paula Creek, Upper Pico).

Pinus muricata D. Don. Axelrod, In R. N. Philbrick, Proc.
   symposium on Biology of California Islands, p. 121, pl. 5,
   fig. 3 (Wilmington, Bixby Slouth).

Pinus muricata D. Don. Langenheim and Durham, Madroño 17:
   38, fig. 3h, 1963 (Little Sur).

The type specimen of P. muricata, as illustrated by Lambert (1837), is presented on pl. 5.

Cones of P. muricata are abundant in the Seacliff collection. Most have suffered compression and are distorted.
They are ovate and asymmetrical, the cones scales are triangular, with most apophyses directed upward. Size is variable
with the larger cones, 8.2 cm. long and 6.0 cm. broad, whereas
the smaller ones are 6.3 cm. long and 4.0 cm. broad. They
are typical of the species as I understand it--with no
variation to symmetrical cones with smooth scales that are
without apophyses, a range included in the species by Linhart
et al. (1967) and by Duffield (1951); the latter are the
result of hybridization-introgression.

Collection: Hypotypes nos. 6272-6274; homeotypes nos.
   6275-6282 7145-7153.

Pinus muricata var. borealis, new var.
(Plates 5 and 6)

Arbor magna etrobusta usque ad 25 m alta; luxuriante
frondosa, foliis perviridibus vel venetis binatis, 9-12 cm
longis; cortice griseo profunde ubique fissurato; strobilis
asymmetricis, ambito plerumque ovatis vel globosis, 4-7 cm
longis 5-7 cm latis, 2-5 verticillitis acute reflecis dense
armatis; apophysibus conicis tumidis sursum spectantibus.

Large robust trees up to 25 m tall; foliage luxuriant,
dark green to bluish-green; needles in two's, 9-12 cm long;
bark dark gray to blackish, heavily and deeply fissured;
cones 4-7 cm long and 5-7 cm broad, in whorls of 2-5,
strongly reflexed, generally ovate to globose, heavily armed
with upswept, swollen (conical) apophyses.

This taxon is found chiefly from Marin County (Inverness
Ridge; San Geronimo Ridge south of Lagunitas) northward,
though relict trees also occur on Huckleberry Hill, Monterey.

Fossil cones previously illustrated from other floras
that represent the var. borealis include:

   Pinus muricata D. Don Mason, Carnegie Inst. Wash. Publ.
   415: 147, pl. 6, fig. 2; pl. 7, fig. 4, 1934 (Millerton).

   Pinus muricata D. Don. Axelrod, In R. N. Philbrick,
   Proc. Symposium on Biology of the California Islands,
   p. 120, pl. 6, figs. 3, 4; pl. 7, figs. 1-9, 1967
   (Point Sal).

   Pinus "borealis" Duffield. Axelrod. Univ. Calif. Publ.
   Geol. Sci. 120: 42, pl. 13, figs. 1-6 and pl. 14, figs.
   1-6 (Carpinteria); pl. 15, figs. 1-6 and pl. 16, figs.
   1-6 (Point Sal), 1980.

Cones of P. muricata var. borealis are not represented in
the collection from Seacliff.

Representative specimens of vars. borealis and stantonii
are deposited at U.C. Davis, Berkeley, LA; Cal. Acad.;
Missouri; Gray; U.S. Nat'l herbaria.

Pinus muricata var. stantonii new var.
(Plates 6, 7 and 11, figs. 1-3)

Arbor parva et humilis 10-20 m alta; aperte frondosa, foliis flavo-virentibus; binatis 8-15 cm longis; cortice cinereo fissurato procatoque; strobilis asymmetricis ambito ovatis vel subtrullatis, 6-7 cm longis 5-5.5 cm latis, 3-5 verticillatiss deflexis; apophyse distale late deltoidea saepe extrorsus spectanti.

Small trees, 10-20 m tall; foliage relatively open, light green; needles in two's 8-15 cm long; bark gray, fissured and ridged; cones asymmetrical, ovate to crudely trulliform in outline, 6-7 cm long and 5-5.5 cm broad, deflexed in whorls of 3-5; cone scales broadly deltoid distally, apophyses directed outward.

It is a pleasure to name this pine for Carey Q. Stanton, previous owner of most of Santa Cruz Island, who for many years has graciously fostered and encouraged biologists and earth scientists to conduct research there.

Four cones in the Seacliff collection comprise the initial fossil record of this pine which is confined now to Santa Cruz Island. The cones are well matched by those produced by the population above Pelican Harbour. The cones are generally oval in outline, the cone scales have broadly triangular apophyses that often are directed outward, and the cones have fewer scales than the associated cones of P. muricata var. muricata. The origin of this taxon is not now clarified by the fossil record. The Pico Formation in which it occurs is separated by the Red Mountain thrust fault from the hills to the north and east which are composed of middle Miocene to Eocene rocks (Jennings and Troxel, 1954). Clasts of these older rocks are in the pebble conglomerate associated with the plant beds, implying derivation from a generally northerly to easterly direction. The hills were much lower than those at present which are the result of Quaternary folding, thrusting, and uplift. Although the evidence indicates that var. stantonii occupied the continental

borderland in the early Pleistocene, it may also have been on
the Channel Islands at that time. Further collections are
needed to clarify its earlier history, and its relationships
to the other varieties of P. muricata.

Collection: Hypotype nos. 7154-7156, homeotype no. 7157.

Pinus radiata D. Don
(Plate 9, figs. 1, 3; pl. 10. fig. 3)

Pinus radiata D. Don. Trans. Linn. Soc. 17: 441, 1836.
  Mason, Carnegie Inst. Wash. Publ. 346: 147, pl. 1, fig.
  2, 1927.
  Chaney and Mason, Carnegie Inst. Wash. Publ. 415: 55, pl.
  4, fig. 2; pl. 5, figs. 4, 9, 1933.
  Axelrod, Univ. Calif. Publ. Geol. Sci. 120: 5, pls. 4-10,
  1980.
Pinus masonii Dorf., Carnegie Inst. Wash. Publ. 412: 70, pl.
  5, fig. 6 only, 1930.

The 24 cones in the collection, most of which are flattened due to compression, are relatively small, averaging 8
cm long. They range from moderately asymmetrical to nearly
symmetrical and are typically ovate-elliptic in outline. The
tips of the scales on the outer side of the asymmetrical
cones are swollen and rounded and without sharply attenuated
apophyses. In some cones, compression has squeezed the
proximal rounded scales sufficiently so that they appear to
be acute. On the more nearly asymmetrical cones, the tips of
the scales are only slightly enlarged.

In size, the suite of cones from Seacliff is closer to
the norm of P. radiata var. binata from Guadalupe Island than
to those produced by the mainland groves at Cambria, Monterey, or Año Nuevo (see Axelrod, 1980a, fig. 2 and pl. 1).
As compared with the Guadalupe Island population, the fossils
have somewhat better-developed apophyses.

Collection: Hypotypes nos. 6283-6285; homeotypes nos.
6286-6288, 7158-7160.

Pinus remorata Mason
(Plate 10, figs. 1, 2)

Pinus remorata Mason, Madrono, 2: 8-10, 1930.

Chaney and Mason, Carnegie Inst. Wash. Publ. 415: 10, pl. 6, figs. 1-5, 1930.

Axelrod, in R. N. Philbrick, Proc. Symposium on Biology of the California Islands, p. 128, pl. 3, fig. 2; pl. 4, figs. 2-4; pl. 5, fig. 6; pl. 8, fig. 3, 1967.

Axelrod, Univ. Calif. Publ. Geol. Sci. 120: 46, pl. 18, figs. 1-5, 1980.

The 4 cones representing this species range from 6.0 to 8.0 cm long and from 4.5 to 5.0 cm broad. They are symmetrically elliptic-ovate in outline, with thin, plane scales which do not show apophyses development on either side. Comparison with suites of cones of the living remorata indicates that they are inseparable.

Similar rare occurrences of P. remorata have been noted at Point Sal and Carpinteria, sites where cones of P. muricata and P. radiata are common to subdominant. The rarity of P. remorata in these deposits appears to reflect its restriction to drier sites, farther removed from the area of plant accumulation in moist, coastal valleys dominated by more mesic conifers.

Collection: Hypotypes nos. 6289-6290; homeotypes nos. 6291, 7161.

Pinus sabiniana Douglas
(Plate 10, fig. 3)

Pinus sabiniana Douglas, ex. D. Don in Lambert, Descr. Genus Pinus, ed. 3, (8°), vol. 2, unnumbered p. between 144-145, pl. 80, 1832.

Douglas, Linn. Soc. London Trans. 16: 749, 1833.

Pinus cf. sabiniana Douglas, Chaney and Mason, Carnegie Inst. Wash. Publ. 415: 56, pl. 5, figs. 7, 8, 1933.

The present record includes an incomplete, flattened cone 11.5 cm long and 7.5 cm wide; its total length is estimated

at 15 cm. The apophyses of the scales are eroded, but the
transverse rhombic shape of the scales and the round-ovate
outline of the cone are typical of P. sabiniana. A second
specimen, measuring 12.5 cm long and 7.5 cm wide represents
the outer portion of a cone. The cone scales have prominent
hooked apophyses. The cones were compared with those of P.
torreyana Parry, but they differ in being broadly ovoid in
outline, the rhombic cone scales are broader than those of
sabiniana or of the fossil, and the apophyses are not so well
developed.

   Collection: Hypotypes nos. 6292, 7162.

Pseudotsuga menziesii (Mirb.) Franco
(Plate 9, fig. 4; Pl. 10, fig. 4)

Pseudotsuga menziesii (Mirb.) Franco, Bol. Soc. Broteriana
   (2) 24: 74, 1950.

Pseudotsuga taxifolia (Lamb.) Britton. Chaney and Mason,
   Carnegie Inst. Wash. Publ. 415: 8, pls. 3, 4 and 5,
   figs. 1-3, 1930.

   Potbury, ibid, 31, pl. 1, fig. 2; pl. 3, fig. 2, 1932.

   Mason, ibid, p. 145, pl. 6, fig. 4; pl. 7, figs. 1, 2,
   1934.

The 6 cones in the collection range from 7.7 to 8.9 cm
long and from 2.5 to 2.7 cm wide. Their slender outline
reflects closure owing to submersion in water. The cones are
similar to those of the living P. menziesii, widely distributed from the outer central Coast Ranges of California northward into British Columbia.

   The cones differ from those of P. macrocarpa (Vasey) Mayr
which is common in the mountains of southern California.
Cones of that species are more oval in outline, the cone
scales are larger, and the overall cone size is generally
more massive.

   Collection: Hypotypes nos. 6293-6294, 7164; homeotypes
nos. 6295-6296, 7165-7167.

CHAPTER II PLATES

(All figures are shown at natural size, except Plate 7.)

PLATE 4

Seacliff Fossil Locality and Allied Modern Vegetation

Fig. 1. View east from Seacliff, Ventura County. Fossil site is just beyond first point of land; Pitas Point in distance. Note Padre Juan Fault, a branch of the Red Mountain Fault system, slicing through the sage- and chaparral-covered hills. (Spence Air Photo, portion of E-3269, taken 11-2-33; from Photo Archives, Dept. of Geography, UCLA).

Fig. 2. Pinus radiata forest in valley of Waddell Creek near Ano Nuevo Point, 25 km northeast of Santa Cruz. Prior to highway construction, the swamp in foreground graded into a lagoon where conifer cones accumulated, thus simulating the concentration of cones at the Seacliff site before their displacement into the marine Pico Basin. Pseudotsuga forest is visible on the ridge and in the valley at the left.

PLATE 5

Type Specimen of Pinus muricata D. Don

Cone (two views) of Pinus muricata D. Don, collected by John M. Coulter and illustrated by A. B. Lambert in A Description of the Genus Pinus, 1928-37 ed., vol. 3, pl. 55. Courtesy of Reed Rollins, Gray Herbarium.

Don's decription states that the pine, collected near San Luis Obispo, was secured at an elevation of 3,000 ft. This is certainly an error, for there are no hills of this height in this region, and the pine nowhere lives at such a level. The actual elevation may have been 300 ft., for P. muricata groves occur at or near this level at several sites in the Pecho Hills west of San Luis Obispo.

Pinus muricata

PLATE 6

Modern Cones of Pinus muricata Varieties

Var. borealis. Note swollen, upswept apophyses and the
compact nature of the mature cone.

Var. muricata. Note that the triangular apophyses are
directed upward.

Var. stantonii. Note the large, broadly deltoid, fewer
cone scales, and the apophyses generally are directed
outward.

var. borealis

var. muricata

var. stantonii

PLATE 7

Modern Cones of Pinus muricata Varieties

Var. borealis.   Typical cones from the Fort Bragg area.

Var. muricata.   Representative cones from La Purisima Ridge, north of Lompoc.

Var. stantonii.   Typical cones from the Pelican Bay population, Santa Cruz Island.

var. borealis

var. muricata

var. stantonii

PLATE 8

Seacliff Fossils

Fig. 1. Pinus attenuata Lemmon. Hypotype no. 6271.

Figs. 2-4. Pinus muricata var. muricata D. Don. Hypotypes nos. 6272-6274.

PLATE 9

Seacliff Fossils

Figs. 1-3. <u>Pinus</u> <u>radiata</u> D. Don.   Hypotypes nos. 6283-6285.

Fig. 4.   <u>Pseudotsuga</u> <u>menziesii</u> (Mirb.) Franco.   Hypotype no. 6294.

PLATE 10

Seacliff Fossils

Figs. 1, 2. Pinus remorata Mason. Hypotypes nos. 6289-6290.

Fig. 3. Pinus sabiniana Douglas. Hypotype no. 6292.

Fig. 4. Pseudotsuga menziesii (Mirb.) Franco. Hypotype no. 6293.

PLATE 11

Seacliff Fossils

Figs. 1-3. Pinus muricata var. stantonii Axelrod. Hypotypes nos. 6254-6256.

Fig. 4. Pseudotsuga menziesii (Mirb.) Franco. Hypotype
no. 7164.

References Cited

AXELROD, D. I.

1937   A Pliocene flora from the Mount Eden beds, southern California. Carnegie Inst. Wash. Publ. 476: 125-183.

1939   A Miocene flora from the western border of the Mohave Desert. Carnegie Inst. Wash. Publ. 516. 128 pp.

1944   The Sonoma flora. Carnegie Inst. Wash. Publ. 553: 167-206.

1950a   A Sonoma florule from Napa. Carnegie Inst. Wash. Publ. 590: 25-71.

1950b   Further studies of the Mount Eden flora, southern California. Carnegie Inst. Wash. Publ. 590: 73-117.

1950c   The Anaverde flora of southern California. Carnegie Inst. Wash. Publ. 590: 119-158.

1950d   The Piru Gorge flora of southern California. Carnegie Inst. Wash. Publ. 590: 161-214.

1958   The Pliocene Verdi flora of western Nevada. Univ. Calif. Publ. Geol. Sci. 34: 91-160.

1966   The Pleistocene Soboba flora of southern California. Univ. Calif. Publ. Geol. Sci. 60. 109 pp.

1967   Evolution of the California closed-cone pine forest. In R. N. Philbrick, ed., Proc. of the symposium on biology of the California islands, pp. 93-101. Santa Barbara Botanic Garden.

1976   History of the coniferous forests, California and
        Nevada. Univ. Calif. Publ. Bot. 70. 62 pp.

1977   Outline history of California vegetation. In
        M. G. Barbour and J. Major, eds., Terrestrial
        vegetation of California, pp. 140-193.
        New York: Wiley.

1978   The origin of coastal sage vegetation, Alta and
        Baja California. Amer. Jour. Bot. 65: 1117-1131.

1979   Age and origin of Sonoran Desert vegetation.
        Calif. Acad. Sci. Occas. Papers 132. 74 pp.

1980a   History of the maritime closed-cone pines, Alta
        and Baja California. Univ. Calif. Publ. Geol.
        Sci. 120. 143 pp.

1980b   Contributions to the Neogene paleobotany of
        central California. Univ. Calif. Publ. Geol. Sci.
        121. 212 pp.

1981   Holocene climatic changes in relation to vegeta-
        tion disjunction and speciation. Amer. Naturalist
        117: 847-870.

1982   Age and origin of the Monterey endemic area.
        Madrono 28: 127-147.

AZZAROLI, A.
1981   About pigmy mammoths of the northern Channel
        Islands and other island faunas. Quaternary
        Research 16: 423-425.

BAILEY, H. P.
1960   A method of determining the warmth and temperate-
        ness of climate. Geografisker Annaler 42: 1-16.

1964   Toward a unified concept of the temperate climate.
        Geog. Rev. 54: 516-545.

BAILEY, T. H. and R. H. JAHNS
1954   Geology of the Transverse Range province, southern
        California. Calif. Div. Mines and Geol. Bull.
        170: chap. 2, pp. 83-106.

BAKER, H. G.
    1972   Seed weight in relation to environmental condi-
            tions in California. Ecology 53: 997-1110.

BALLANCE, P. F., M. R. GREGORY, and G. W. GIBSON
    1981   Coconuts in Miocene turbidites in New Zealand:
            possible evidence for tsunami origin of some
            turbidite currents. Geology 9: 592-595.

BERGER, R. and W. F. LIBBY
    1966   UCLA Radiocarbon dates V. Radiocarbon 8: 467-497.

BOELLSTORFF, J.
    1978   North American Pleistocene Stages reconsidered
            in the light of the probable Pliocene-Pleistocene
            continental glaciation. Science 202: 305-307.

BOBROV, E. G.
    1983   Introgressive hybridization and geohistorical
            changes of formations in the taiga zone of the
            USSR. Bot. Zurhnal 68(1): 3-9.

CAMPBELL, C. A.
    1974   Paleoecological analysis of molluscan assemblages
            from the Pleistocene Millerton Formation. M.S.
            thesis, Univ. California, Davis. 114 pp.

CHANEY, R. W., and H. L. MASON
    1930   A Pleistocene flora from Santa Cruz Island,
            California. Carnegie Inst. Wash. Publ. 415: 1-24.

    1933   A Pleistocene flora from the asphalt deposits at
            Carpinteria, California. Carnegie Inst. Wash.
            Publ. 415: 47-77.

CHRISTENSEN, M. N.
    1966   The Quaternary of the California Coast Ranges.
            Calif. Div. Mines and Geol. Bull. 190: 305-314.

CONDIT, C.
    1944   The Table Mountain flora. Carnegie Inst. Wash.
            Publ. 553: 57-90.

COOK, J.
    1978   Soil survey of Monterey County, California. USDA
            Soil Conservation Service. 228 pp., 188 map
            sheets.

CRITCHFIELD, W. B.
  1967   Crossability and relationships of the closed-cone
          pines. Silvae Genetica 16: 89-97.
CROWELL, J. C.
  1967   Origin of pebbly mudstones. Geol. Soc. Amer.
          Bull. 68: 993-1010.
CURRY, R. R.
  1966   Glaciation about 3,000,000 years ago in the Sierra
          Nevada. Science 154: 770-771.
  1971   Glacial and Pleistocene history of the Mammoth
          Lakes Sierra, California. Univ. Montana Dept.
          Geology, Geological Series Publ. 11. 49 pp.
DALRYMPLE, G. B.
  1963   Potassium-argon dates of some Cenozoic volcanic
          rocks of the Sierra Nevada, California. Geol.
          Soc. Amer. Bull. 74: 387-390.
DIBBLEE, T. W., JR.
  1966   Geology of the central Santa Ynez Mountains, Santa
          Barbara County, California. Calif. Div. Mines and
          Geol. Bull. 186. 99 pp.
DON, D.
  1836   Description of five new species of the genus Pinus,
          discovered by Dr. Coulter in California. Linn.
          Soc. London Trans. 17: 441.
DORF, E.
  1930   Pliocene floras of California. Carnegie Inst.
          Wash. Publ. 412: 1-112.
DUFFIELD, J. W.
  1951   Interrelationships of the California closed-cone
          pines with special reference to Pinus muricata D.
          Don. Ph. D. thesis, Univ. California, Berkeley.
          77 pp.
EVERNDEN, J. F., and R. K. S. EVERNDEN
  1970   The Cenozoic time scale. Geol. Soc. Amer. Spec.
          Paper 124: 71-90.

EVERNDEN, J. F. and G. T. JAMES
    1964   Potassium-argon dates and the Tertiary floras of
           North America. Amer. Jour. Sci. 262: 945-974.

FERGUSSON, G. J. and W. F. LIBBY
    1966   UCLA radiocarbon dates. II. Radiocarbon 5: 1-22.

FIELDING, J. M.
    1953   Variation in Monterey pine. Forest Timber Bur.
           Bull. 31: 1-43.

FLINT, R. F.
    1971   Glacial and Quaternary geology. New York: Wiley;
           892 pp.

FRICK, C.
    1921   Extinct vertebrate faunas of the badlands of
           Bautista Creek and San Timote Canyon, southern
           California. Univ. Calif. Publ. Geol. Sci. 12:
           277-424.

GALLOWAY, A. J.
    1977   Geology of the Point Reyes Peninsula, Marin Co.,
           California. Calif. Div. Mines and Geol. Bull.
           202: 1-72.

GARDNER, A., JR., and K. E. BRADSHAW
    1954   Characteristics and vegetation relationships of
           some podsolic soils near the coast of northern
           California. Soil Soc. Amer. Proc. 18: 320-325.

GLEN, W.
    1959   Pliocene and Lower Pleistocene of the western part
           of the San Francisco Peninsula. Univ. Calif.
           Publ. Geol. Sci. 36: 147-198.

GRIFFIN, J. R., and W. B. CRITCHFIELD
    1972   The distribution of forest trees in California.
           USDA Forest Serv. Research paper PSW 82/1972.
           114 pp.

HELLEY, E. J., D. P. ADAM, and D. B. BURKE

    1972   Late Quaternary and stratigraphic and paleo-
            ecological investigations in the San Francisco
            Bay area. <u>In</u> V. A. Frizzel, ed., Progress report
            on the USGS Quaternary studies in the San Fran-
            cisco Bay area, pp. 19-29. Guidebook for Friends
            of the Pleistocene, Oct. 6-8, 1972.

HELLEY, E. J. and K. R. LAJOIE

    1979   Flatland deposits of the San Francisco Bay region,
            California--their geology and engineering proper-
            ties and their importance in comprehensive plan-
            ning. U.S. Geol. Surv. Prof. Paper 943. 88 pp.

HEUSSER, L. E.

    1978   Pollen in Santa Barbara basin, California: a
            12,000 year record. Geol. Soc. Amer. Bull. 89:
            673-678.

HIBBARD, C. W., D. E. RAY, D. E. SAVAGE, D. W. TAYLOR, and
    J. E. GUILDAY

    1965   Quaternary mammals of North America. <u>In</u> H. E.
            Wright, Jr. and D. E. Frey, eds., The Quaternary
            of the United States, pp. 509-525. Princeton,
            N. J., Princeton Univ. Press.

HUBER, N. K.

    1981   Amount and timing of Late Cenozoic uplift and tilt
            of the central Sierra Nevada, California--evidence
            from the upper San Joaquin River basin. U.S.
            Geol. Surv. Prof. Paper 1197. 20 pp.

IZETT, G. A., C. W. NAESER, and J. D. OBRADOVICH

    1974   Fission track age of zircons from an ash bed in
            the Pico Formation (Pliocene and Pleistocene) near
            Ventura, California. Geol. Soc. Amer. Abstracts
            with Programs, 6(3): 197.

JENNINGS, C. W. and R. G. STRAND

    1958   Geologic Map of California: Santa Cruz sheet.
            Calif. Div. Mines and Geol. (1964 ed.)

    1969   Geologic Map of California. Los Angeles sheet.
            Calif. Div. Mines and Geol.

JENNINGS, C. W., and B. W. TROXEL
    1954   Ventura Basin. Geologic Guide No. 2, 63 pp. In
             R. H. Jahns, ed., Geology of southern California,
             Calif. Div. Mines and Geol. Bull. 170.

JENNY, H., R. J. ARKLEY, and A. M. SCHULTZ
    1961   Pygmy forest-podsol ecosystem and its dune
             associates of the Mendocino coast. Madroño
             20: 60-74.

LAMBERT, A. B.
    1837   A description of the genus Pinus, vol. 3. London:
             Geo. White.

LANGENHEIM, J. Y., and J. W. DURHAM
    1963   Quaternary closed-cone pine forest from travertine
             near Little Sur, California. Madroño 17: 33-51.

LEIGHLY, J.
    1938   The extremes of annual temperature march with
             particular reference to California. Univ. Calif.
             Publ. Geog. 6: 191-234.

LEMMON, J. G.
    1895   West American cone-bearers, 3rd ed. Oakland.
             104 pp.

LINHART, Y. B., B. BURR, and M. T. CONKLE
    1967   The closed-cone pines of the northern Channel
             Islands. In R. N. Philbrick, ed., Proc. sympos-
             ium on the biology of the California Islands, pp.
             151-177. Santa Barbara Botanic Garden.

LIPMAN, P. W., H. J. PROTSKA, and R. L. CHRISTIANSEN
    1972   Cenozoic volcanism and plate tectonic evolution of
             the western United States. I. Early and Middle
             Cenozoic. Roy. Soc. London Philos. Trans., Ser.
             A, 271: 217-248.

MACGINITIE, H. D.
    1953   Fossil plants of the Florissant beds, Colorado.
             Carnegie Inst. Wash. Publ. 599. 188 pp.
    1969   The Eocene Green River flora of northwestern
             Colorado and northeastern Utah. Univ. Calif.
             Publ. Geol. Sci. 83. 140 pp.

MARSHALL, L. G., R. B. BUTLER, R. E. DRAKE, and G. H. CURTIS
   1982   Geochronology of the type Uquian (Late Cenozoic)
          land mammal age, Argentina. Science 216: 986-989.
MARTIN, P. S. and J. S. GUILDAY
   1967   A bestiary for Pleistocene biologists. In P. S.
          Martin and H. E. Wright, Jr., eds., Pleistocene
          extinctions: The search for a cause, pp. 1-62.
          New Haven, Conn. Yale Univ. Press.
MASON, H. L.
   1927   Fossil records of some west American conifers.
          Carnegie Inst. Wash. Publ. 346: 139-158.

   1932   A phylogenetic series of the California closed-
          cone pines suggested by the fossil record.
          Madroño 2: 49-55.

   1934   Pleistocene flora of the Tomales Formation.
          Carnegie Inst. Wash. Publ. 415: 81-197.

   1949   Evidence for the genetic submergence of Pinus
          remorata. In G. L. Jepsen, G. G. Simpson, and E.
          Mayr, eds., Genetics, paleontology and evolution,
          pp. 356-362. Princeton, N. J.: Princeton Univ.
          Press.
MATHIESON, S. A. and A. M. SARNA-WOJCICKI
   1982   Ash layer in Mohawk Valley, Plumas County, Cali-
          fornia, correlated with the 0.45-m.y.-old Rockland
          ash--implications for the glacial and lacustrine
          history of the region. Geol. Soc. Amer. Abstracts
          with Programs 14(4): 184.
MATTI, J. C., and D. M. MORTON
   1975   Geologic history of the San Timoteo badlands,
          southern California. Geol. Soc. Amer. Abstracts
          with Programs 7(3): 344.
MAYR, H.
   1890   Die Waldungen von Nordamerika. Munich: M.
          Reiger. 488 pp.

MCDOWELL, F. W. AND S. E. CLABAUGH
  1979   Ignimbrites of the Sierra Madre Occidental and
         their relation to the tectonic history of western
         Mexico. Geol. Soc. Amer. Spec. Paper 180: 113-124.

METCALF, W.
  1923   An ancient pine cone. Amer. Forestry 29: 172.

MURPHY, T. M.
  1981   Immunochemical comparisons of seed proteins from
         populations of Pinus radiata (Pinaceae). Amer.
         Jour. Bot. 68: 254-269.

NEWCOMB, G. B.
  1962   Geographic variation in Pinus attenuata Lemmon.
         Ph.D. thesis, Univ. California, Berkeley. 191 p.

NOBLE, D. C., D. B. SLEMMONS, M. J. KORRINGA, W. R. DICKIN-
         SON, U. AL-RAWI, and E. H. MCKEE
  1974   Eureka Valley Tuff, east central California and
         adjacent Nevada. Geology 2: 139-142.

POTBURY, S. S.
  1932   A Pleistocene flora from San Bruno, San Mateo
         County, California. Carnegie Inst. Wash. Publ.
         415: 25-44.

PUTNAM, W. C.
  1962   Late Cenozoic geology of McGee Mountain, Mono
         County, California. Univ. Calif. Publ. Geol. Sci.
         40: 181-218.

RAVEN, P. H., and D. I. AXELROD
  1978   Origin and relationships of the California flora.
         Univ. Calif. Publ. Botany 72. 134 pp.

SARNA-WOJCICKI, A. M.
  1976   Correlation of late Cenozoic tuffs in the central
         Coast Ranges of California by means of trace- and
         minor-element chemistry. U.S. Geol. Surv. Prof.
         Paper 972. 33 pp.

SARNA-WOJCICKI, A. M.

    1979   Chemical correlation of some late Cenozoic tuffs
           of northern and central California by neutron
           activation analysis of glass and comparison with
           X-ray fluorescence analysis. U.S. Geol. Surv.
           Prof. Paper 1147. 15 pp.

SAVAGE, D. E.

    1951   Late Cenozoic vertebrates of the San Francisco Bay
           region. Univ. Calif. Publ. Geol. Sci. 28: 215-
           314.

SHAW, G. R.

    1914   The genus *Pinus*. Arnold Arboretum Publ. 5.
           95 pp.

THOMAS, J. H.

    1961   Flora of the Santa Cruz Mountains of California.
           Stanford, Calif.: Stanford Univ. Press. 434 pp.

TING, W. S.

    1966   Determination of *Pinus* species by pollen statis-
           tics. Univ. Calif. Pub. Geol. Sci. 58. 182 pp.

WARTER, J. K.

    1976   Late Pleistocene plant communities--evidence from
           the Rancho La Brea tar pits. *In* J. Latting, ed.,
           Plant communities of southern California. Calif.
           Native Plant Soc. Spec. Publ. 2: 32-39.

WEBBER, I. E.

    1933   Woods from the Ricardo Pliocene of Last Chance
           Gulch, California. Carnegie Inst. Wash. Pub.
           412: 113-134.

WIGGINS, I. L.

    1951   An additional specimen of *Pinus pieperi* Dorf from
           Ventura County, California. Amer. Jour. Bot. 38:
           211-213.

YEATS, R. S., M. N. CLARK, E. A. KELLER, and T. K. ROCKWELL

    1981   Active fault hazard in southern California:
           ground rupture versus seismic shaking. Geol. Soc.
           Amer. Bull. 92: 189-196.